AF545057

Data Sharing for International Water Resource Management:
Eastern Europe, Russia and the CIS

NATO Science Series

A Series presenting the results of activities sponsored by the NATO Science Committee. The Series is published by IOS Press and Kluwer Academic Publishers, in conjunction with the NATO Scientific Affairs Division.

A. Life Sciences	IOS Press
B. Physics	Kluwer Academic Publishers
C. Mathematical and Physical Sciences	Kluwer Academic Publishers
D. Behavioural and Social Sciences	Kluwer Academic Publishers
E. Applied Sciences	Kluwer Academic Publishers
F. Computer and Systems Sciences	IOS Press

1. Disarmament Technologies	Kluwer Academic Publishers
2. Environmental Security	Kluwer Academic Publishers
3. High Technology	Kluwer Academic Publishers
4. Science and Technology Policy	IOS Press
5. Computer Networking	IOS Press

NATO-PCO-DATA BASE

The NATO Science Series continues the series of books published formerly in the NATO ASI Series. An electronic index to the NATO ASI Series provides full bibliographical references (with keywords and/or abstracts) to more than 50000 contributions from internatonal scientists published in all sections of the NATO ASI Series.
Access to the NATO-PCO-DATA BASE is possible via CD-ROM "NATO-PCO-DATA BASE" with user-friendly retrieval software in English, French and German (WTV GmbH and DATAWARE Technologies Inc. 1989).

The CD-ROM of the NATO ASI Series can be ordered from: PCO, Overijse, Belgium

2. Environmental Security – Vol. 61

Data Sharing for International Water Resource Management: Eastern Europe, Russia and the CIS

edited by

Thomas Naff

Department of Asian and Middle Eastern Studies and
Center of Earth and Environmental Science,
University of Pennsylvania,
Philadelphia, U.S.A.

Kluwer Academic Publishers

Dordrecht / Boston / London

Published in cooperation with NATO Scientific Affairs Division

Proceedings of the NATO Advanced Research Workshop on
Managing Environmental Degradation: Dialogue, Cooperation, Data Sharing
Budapest, Hungary
27–31 August 1997

A C.I.P. Catalogue record for this book is available from the Library of Congress.

ISBN 0-7923-5917-8

Published by Kluwer Academic Publishers,
P.O. Box 17, 3300 AA Dordrecht, The Netherlands.

Sold and distributed in North, Central and South America
by Kluwer Academic Publishers,
101 Philip Drive, Norwell, MA 02061, U.S.A.

In all other countries, sold and distributed
by Kluwer Academic Publishers,
P.O. Box 322, 3300 AH Dordrecht, The Netherlands.

Printed on acid-free paper

Printed in the Netherlands.

TABLE OF CONTENTS

PREFACE ~ ACKNOWLEDGEMENTS

This volume of papers resulted from a NATO Advanced Research Workshop (ARW) which brought together in Budapest, Hungary August 27 - 31, 1997 a company of specialists on environmental and water resource management from the United States, Eastern Europe, Russia, and the CIS (Commonwealth of Independent States — formerly parts of the Soviet Union), and from several international agencies. This ARW was one of an ongoing series on environmental security, an issue designated by the NATO Science Committee as a priority area of study. The ARWs are intended to convene in dialogue and collaboration members of the scientific and technological communities of NATO member nations and those nations specified by NATO as Cooperation Partner countries (CPCs).*

Under these joint scientific endeavors, such urgent problems as environmental and human health, nuclear and chemical pollution, water degradation and water resource management, air pollution, natural and man-made hazards, climate change, agricultural management, and more, are examined with a view to recommending amelioration and practical solutions and to identifying effective in-country capacity-building measures. In undertaking such an enlightened program, the NATO Science Committee has not only acknowledged that the environment is a critical determinant of security and therefore the latter concept must be redefined accordingly, but it has led the way in illuminating the issues.

The success of our ARW was due in no small measure to the support and guidance of Dr. L. Veiga da Cunha, Director of the Priority Area on Environmental Security, NATO Scientific and Environmental Affairs Division. In addition, the cooperation and assistance of Dr. Akos R. Herman, Co-Director of the ARW and Director of the National Technical Information Centre and Library, Budapest, and his most able assistant, Julia Fodor were indispensable. Dr. Herman not only made the Library, its facilities and staff available, but together with Ms. Fodor faced and solved the inevitable bureaucratic, human, and logistical problems with patience, humor, kindness, and ingenuity. Professor Evan Vlachos, Associate Director, International School for Water Resources, Colorado State University, Boulder Colorado, lent his wisdom and experience at critical moments. Finally, my talented student assistants, Rachel Brown, Tamara Galinsky, Katherine Glassmyer, and Camille Kellett, who manned computers, telephones, and fax machines, interpreted, took notes, kept track of countless pieces of paper and the matters inscribed thereon, on both sides of the Atlantic, before and after the workshop, must be accorded my personal special thanks, particularly to Camille Kellett whose technical skills and intelligent eye have been essential in the final preparation of this book.

Thomas Naff
Professor
University of Pennsylvania
Philadelphia
April 1999

*CPCs: Albania, Armenia, Azerbaijan, Belarus, Bulgaria, Czech Republic, Estonia, Georgia, Hungary, Kazakhstan, Kyrgystan, Latvia, Lithuania, Moldova, Poland, Russian Federation, Slovak Republic, Tajikistan, Turkmenistan, Ukraine, Uzbekistan

INTRODUCTION

The Issues, Conclusions, and Recommendations of the NATO Advanced Research Workshop — Budapest, Hungary July 27 - 31, 1997

THOMAS NAFF
University of Pennsylvania
847 Williams Hall
Philadelphia, PA 19104-6305 USA
tnaff@sas.upenn.edu

1. The Issues

Sharing data and information enables people to think together in solving problems, in building trust essential for cooperative efforts toward sustaining shared vital natural resources, and in avoiding conflict. It is axiomatic that all planning and policy making, not least for environmental and resource sustainability, depend for success on accurate data and information dispensed freely to all who need it, from farmers to heads of state. These maxims are particularly apt when applied to water resources that are international and transboundary. In those circumstances, the need for cooperation and sharing are acute if the water source is to be managed, distributed, and used equitably and efficiently.

In many parts of the world, the collection, management, reporting, and quality of water and environmental data are often so poor and incomplete as to render them useless, or they are treated as security issues and are therefore classified. Either way, wherever those conditions exist, essential planning and policy data and information of good quality are relatively hard to come by. The consequences are high, particularly for effective basin-wide river management and resource sustainability. Paradoxically, this state of affairs obtains at a time when new, rapidly evolving technologies involving remote sensing, satellites, computers, and the World-Wide Web enable us to gather and distribute data in ways, on a scale, and with a speed heretofore unimagined.

Even when there are no serious obstacles to cooperation among the users of a multinational waterway, the relationship between water management and data remains troubled, as one of the essays in this volume reveals. The reasons are several and obvious: fresh waterways are dynamic systems and the information that is carried within the flow of water is often difficult and expensive to measure; rivers are used sequentially in ways that reflect the various — sometimes contradictory — interests of the sovereign riparians along the river basin; and when the information is not well gathered or efficiently processed and shared among those who need it, the problems of management and planning are made more difficult.

2. The NATO ARW

A NATO Advanced Research Workshop (ARW), entitled "Data Sharing for International Water Resource Management: Eastern Europe, Russia, and the CIS," was convened from July 27 - 31, 1997 at the National Technical Information Centre and Library (OMIKK), in Budapest for the purpose of exploring ways to deal with these and related problems.

2.1. PARTICIPANTS

Workshop members represented thirteen nations and five international organizations (see the roster of individual participants and their affiliations). They were: Albania, Belarus, Hungary, Kazakhstan, Kyrgystan, Netherlands, Poland, Russia, The Slovak Republic, Turkmenistan, Ukraine, the United States and Uzbekistan. The international organizations represented were FAO (The Food and Agricultural Organization of the United Nations, Rome), DANREG (A cooperative agency of Danube riparians Austria, Hungary, and the Slovak Republic), REC (Regional and Environmental Center, Central and Eastern Europe), UNDP and the World Bank. The twelve represented nations other than the US are designated by NATO as Cooperation Partner countries (CPCs). CPCs are non-NATO countries that cooperate with NATO in various activities. All of the workshop participants, except those representing international organizations, came from a CPC or NATO country.

2.2. OBJECTIVES

The major objective set by the workshop members was that an effort should be made to go beyond simply reporting and discussing issues to promoting the acceptance of the basic data-sharing propositions that shared data is the most productive and efficiently used data, that the ideal of peaceful, cooperative use of water and the environment cannot be achieved in the absence of sharing, and that sharing is the foundation of good neighborliness.

These precepts became the underlying tenets of the other five, inter-related workshop goals: 1) to demonstrate the central need for data sharing and distributed database networks for effective national and regional resource management; 2) to identify the most appropriate and cost effective technologies and systems — and their applications — required for efficient data gathering, processing, monitoring, storage, and distribution; 3) to examine capacity building, costs, and funding sources; 4) to identify and analyze non-technical barriers to data sharing and distribution via networks — e.g., bureaucratic, administrative, political, attitudinal, etc; 5) to develop realistic, achievable recommendations for post-workshop actions by NATO and other international organizations that will assist each represented nation to attain the agreed-upon goals.

3. Workshop Conclusions and Recommendations

The workshop was organized into two parts. The first consisted of the presentations and discussions of papers on the theme of the workshop by key speakers and participants. The principal activity of the second half of the workshop was the deliberations of three separate working groups, into which the members of the workshop were divided. The task of each of the working groups was to consider the major problems raised in the papers and discussions, examine various means and applications to solve or alleviate them, and to recommend practical plans of action. The major issues considered by the working groups were divided into three related topics: technological needs, training and skilled manpower needs, and incentives for effective data collection and data sharing, particularly as they relate to users. The issues of costs and funding were integrated into each of the working group topics.

4. General Conclusions

- Given the magnitude, complexity, and severity of the environmental and water problems faced by the CPC nations, and the limited financial, technical, and manpower resources they possess to solve them, immediate mitigating actions with financial, technical, and training assistance from the international community is essential if potentially serious crises are to be avoided.

- The solutions must be commensurate with the complexity of the problems and should be designed for both short and long term effectiveness.

- It is recognized that the process of reversing the consequences of decades of environmental and water resource degradation requires from beginning to end the effective collection, management, distribution, and sharing of accurate data and information. This objective would be achieved best through coordinated distributed database networks.

- Because the major problems are transboundary and international, they are most productively attacked cooperatively, basin-wide and region-wide. Therefore, data must be made accessible to and shared by all those end-users who need it. Without data distribution and sharing, attempted solutions would be neither efficient nor cost-effective.

- Although there presently exist relatively good data and knowledge about the environmental and water problems within the CPCs, it tends to be fragmented with critical gaps that need to be filled; there is also a need for updated knowledge in data-gathering technology, training in the application of that technology, and a means for staying abreast of technological developments.

- While there is good expertise in the relevant fields within the CPCs, that expertise appears to be growing thin in relation to the increasing size of the problems and needs enhancement. The consequence of these circumstances is that there is insufficient uniformity in the processes of data collection, quality monitoring, and standardization of measurements and techniques, which seriously reduces essential comparability of data. These conditions severely handicap efforts of CPCs to solve growing environmental and water problems, most of which are regional.

- With some relatively modest and timely assistance now, the CPCs have, with a few exceptions, the indigenous capacity to deal with their problems, and they understand the necessity for regional cooperation. They understand too that without swift, effective action, grim, destabilizing, environmental and water-related crises loom on the short horizon.

- To achieve the general post-workshop objectives and to maintain their benefits, it would be necessary to promote among policy and decision makers, technocrats, and the public the advantages to be gained from the sustained collection of good data and their distribution and sharing through good information services.

5. Paramount Recommendations

Collectively, the ARW working groups proposed altogether eleven recommendations for action, but four were put forward as paramount. They are perceived to be practical building blocks to other productive actions, to be affordable, achievable, and the ones that would yield the quickest and most beneficial results:

- Plan and develop programs and projects that would demonstrate to policy formulators, decision-makers, technocrats, and the public the advantages of continuous quality data collection and good information services together with distribution and sharing of that data to all end-users.

- Plan and develop as soon as possible a small, three-year, three-phase project for the creation of a pilot distributed database network located in one of the regions covered by the workshop, which would, by the end of the final phase, extend significantly the range of data collected and shared and would include all regional actors who wish to participate.

- Plan and develop as early as possible for CPCs a series of intensive, short (six to eight week) training courses on data technology, collection, management, and distribution modeled on those courses offered by such organizations as the US Geological Survey and specialized agencies of the UN.

- Plan and develop as early as possible a series of short seminars for CPC representatives on how to prepare project proposals and budgets for submission to foundations, international organizations, and other funding sources in Western Europe, North America, and Japan. These seminars would be organized with the assistance of representatives of foundations, local Non-Governmental Organizations (NGOs) and other funding agencies.

6. Other Related Recommendations

The remaining workshop recommendations are offered in an order of priority determined by each working group:

6.1. WORKING GROUP I: TECHNOLOGICAL NEEDS

- Organize a workshop for the development of uniform methods of ecological monitoring of water systems and for the timely sharing of the resulting data among end-users.

- Organize a workshop on the topic "The Role of Database Systems in the Science, Economics, Sociology, and Environmental Protection Associated with Flood Disasters."

- Organize a project for the purpose of defining environmental assessment standards for hydrological development with a view to making it possible to collect such data uniformly.

- Organize a workshop on the applications of Internet technologies by hydrological services and institutes in the regions covered by the workshop.

6.2. WORKING GROUP II: SKILLED MANPOWER NEEDS AND TRAINING

- Plan and develop as early as possible a series of intensive, short (six to eight week) training courses on data technology, collection, management, and distribution modeled on those courses offered by such organizations as the US Geological Survey and specialized agencies of the UN. (N.B. It will be noted that this recommendation is the third item under the heading Paramount Recommendations.)

- Seek a NATO Linkage Grant for planning the establishment of a Regional Training Center that involves Cooperation Partner countries and NATO member countries. The purpose of the center would be to provide training in basic and advanced skills in the earth sciences. Integral to the center's mission would be training in the collection, management, and distribution of data relating to earth sciences. The Center would collect data and be repository of data, and could coordinate a regional distributed database network.

6.3. WORKING GROUP III: INCENTIVES FOR DATA COLLECTION AND SHARING IN RELATION TO END-USERS

- By means of workshops and institutes, train technical staff in effective techniques for the marketing of data and information services with a view to applying the acquired skills to various outreach activities.

- Promote the exchange of data on both international and domestic levels by the provision of grants to selected institutions in CPCs for the following related activities: the creation and maintenance of widely accessible Internet Homepages; the compilation and distribution of a standardized glossary of terms used in information systems which is cross-referenced in several languages; construct a meta-database clearinghouse for the assistance of both data providers and users and as a means of keeping them efficiently linked.

- Plan and develop programs and projects that would demonstrate to policy formulators, decision makers, technocrats, and the public the advantages of continuous quality data collection and good information services together with the distribution and sharing of that data to all end-users. (N.B. This is the first item under Paramount Recommendations.)

The members further agreed on the underlying principle that science and other forms of scholarship, international business, social and economic development, and national capacity building cannot be advanced in today's complicated, interdependent world without freely shared knowledge, information, and data among individuals, institutions, and organizations. The inspiration for these conclusions and recommendations are clearly reflected in the contributions to this volume by the members of the workshop.

A CASE FOR DATA SHARING AND DATABASE NETWORKS IN THE MANAGEMENT OF WATER RESOURCES AND THE ENVIRONMENT

THOMAS NAFF
University of Pennsylvania
847 Williams Hall
Philadelphia, PA 19104-6305 USA
tnaff@sas.upenn.edu

Abstract

Solutions to water-related problems must begin and end with good information and data. The gathering of data, their quality, comparability, and management, who controls and distributes them, how they are used, by whom, and toward what ends are potent determinants of the well-being of individuals and nations. Flowing water carries within itself its own information, and because watercourses tend to be very dynamic systems, water data are consequently almost always in one degree or another estimates, extrapolations, and, sometimes, even speculation. This study analyzes these issues and emphasizes the importance of cooperative data collection, the sharing of data, and argues the necessity for establishing a nexus of distributed water information systems, offering a specific model based on new relational software.

Key Words

comparability, data, database, discharge, GIS, information, management, measurements, networks, pointer system, quality, sharing

1. Introduction: Water and Information

Information and water are alike in many ways: both are pervasive, salient, intense, aggregated, complicated, and rarely (if ever) neutral. Like water, information flows until it is somehow dammed up or runs out, it can be both beneficent or deadly in its combinations and uses, and information confers power which inspires a strong tendency to own or control it. Possession of information in its various forms can link individuals to the mainstream of knowledge and culture or its absence may relegate them to lives of ignorance and toil for little gain and makes the imposition and maintenance of authoritarian regimes all the more possible.

T. Naff (ed.),
Data Sharing for International Water Resource Management: Eastern Europe, Russia and the CIS, 1–11.

2. Reasons for Data Sharing

The justifications for data sharing should be self-evident, but are all too often subordinated to political ideology and misperceived expediencies. It thus sometimes serves a useful, if not salutary, purpose to reiterate some the underlying propositions:

- Sharing data and information enables people to think together in solving problems, to build trust, and avoid conflict.
- Sharing is the foundation of good neighborliness.
- The ideal of peaceful, cooperative use of water and the environment cannot be achieved in the absence of sharing.
- Shared data is the most productive and efficiently used data.
- All planning and policy making, not least for resource and environmental sustainability, depend for success on accurate data and information distributed freely to all who need it.
- Science and other forms of scholarship, international business, social and economic development, and national capacity building, cannot be advanced in today's complicated, interdependent world without freely shared knowledge, information, and data among individuals, institutions, and organizations.
- The issues of gathering information, its quality, its management, who controls and distributes it, how it is used, by whom, and toward what ends are potent determinants of the well-being of individuals and nations.
- Solutions to water problems must be commensurate with their complexity; this is not possible without adequate, shared data.
- The latter proposition also applies if there is to be equitable distribution of water among the users of a transnational water source.
- Cooperative data collection and sharing prevents unnecessary duplication and reduces costs for all involved parties.
- Information is already among the world's most vital resources; it is at the heart of economic and social development, it creates wealth, and will only increase in importance.

The worldwide revolution in electronic communications has produced an explosion of information dazzling in its variety and accessible on the Internet and World Wide Web. It is a technological revolution that has reduced our planet to a virtual global village whose remotest parts will, in the not-too-distant future be electronically connected. The continual introduction of sophisticated innovations in computer-based telecommunications and information-gathering technology make efforts to avoid sharing certain kinds of water-resource and environmental information increasingly difficult, senseless, and even self-defeating.

3. Water Management and Data: A Troubled Relationship

There is widespread agreement among experts that effective solutions to water-related and environmental problems begin and end with good information and data. Despite expert consensus on these issues, the relationship between water management and data remains troubled. There are several obvious reasons: fresh waterways are dynamic systems; rivers are used sequentially in ways that reflect the various—sometimes contradictory—interests of the sovereign riparians along river basin (the most complex of these interests involve issues of control, distribution, and use); the information that is carried within the flow of water is often difficult and expensive to measure; and when the information is not well gathered or efficiently processed and shared among those who need it, the problems of management and planning are made more difficult. Much of what is true of water applies to the environment as well.

4. The Nature of Water Data and Analysis

Data Measurements. As water issues are very complex it follows that water data will always mirror that complexity. Such data are generated from a wide variety of measurements and the degree of accuracy is determined by an equally wide compass of factors, e.g.: how data are defined; the qualifications and motives of the data gatherers; how the data are gathered—i.e., the technique and system of measurement; the nature, place, and time of measurement; field efficiency; the purpose of measurement; the cultural context within which the data are collected; and prevailing ecological and environmental factors.

The Quality of Data. The quality of data is as important as its quantity. Good baseline data are essential for analysis of all hydrological issues and for the formulation of sound water policy; this requirement makes the historical record very important, but the latter data are often inconsistent or inaccurate owing to the problems of collection, measurement, and processing—and, of course, to the natural changes or those caused by humans to any body of water over time. At all events, a full range of data are needed to inscribe the multiple uses of water. The sharing of stream data is essential for achieving the full potential of any international basin system or combination of systems.

Use of Water Data. Corollary to the problems of data complexity and collection, are those encountered in using water data. Problems of use are numerous, caused by such factors as natural variations in climate and the flows of river systems; irregular, unsystematic collection; little or no compilation and distribution of data; inconsistent standards of measurement, which is a function of the absence of a uniform standard for international units of measurement; imprecise, divergent terminology and definitions; widely varying skills and experience among the collectors of information, which produces poor quality data; and ulterior motivations of some data gatherers and the ideological or political uses to which the data are put.

Discharge Measurements. Further compounding the problems of data gathering is the fact that essential discharge measurements of large rivers and estuaries is difficult, time-consuming, and sometimes even dangerous. Often, standard discharge measuring techniques (e.g., ultra-velocity or ultrasonic velocity metering) cannot be made in tide-affected rivers and estuaries owing to highly dynamic discharge conditions [1]. (The definition of "discharge" as used herein is the volume of water and all matter dissolved in or mixed with it that that flows past a fixed point in a given interval of time.)

Types of Data. Beyond such problems are questions of types of data: measured, derived, or computed, which are acquired, respectively, by direct field measurements or remote sensing, by deduction from other available data, or by using computers to calculate national or large-scale figures based on assumptions or approximations drawn from established usage patterns. These types of data gathering have their own limitations; for example, direct field collection is expensive and usually politically sensitive while derived and computed data can be only as good as the input data and assumptions on which they are based [2]. The upshot of these issues for planners and policy makers is that only short-term projections can be made with any degree of accuracy.

5. Comparability: A Key Issue of Water Data Collection

Another key issue of water data, however they are collected, is comparability. The production of data that are comparable is essential to minimizing the possibility of false conclusions; but variations in method and competence render this matter problematic. The same data may be collected simultaneously on the same body of water on either side of a political border, or on opposite banks of a stream with the same set of objectives, but if the methods employed are so different as to be incompatible, the results will not be comparable.

The problem of comparability is not confined to accumulations of in-stream data but extends to the laboratory analyses of the collected water. For example, a water quality laboratory may have abandoned a given technique for estimating a dissolved substance in water and replaced it with another method, perhaps in 1980. When analysts later examine the water quality data produced by the earlier technique, say for the period 1970 - 1990, they might observe a step-function in the dissolved constituent leading them to assume that the changes therein occurred in nature rather than in the laboratory because of an incomparability between methodologically induced changes and changes that occur naturally. Once errors of assumption or data rooted in the problem of comparability get into the scientific or technical literature, it is often difficult to correct them for years, and their impact can be cumulative.

6. Need: The Importance of Quantitative and Quality Data

Rivers and aquifers are monitored for the quantity and quality of their water. Quantity data are used principally for planning and operations while quality data are necessary for the maintenance of human and aquatic health [3]. There are a wide range of needs for such data. The following typology of some 19 compound reasons why real-time hydrometeorological data are necessary illustrate the point. It is instructive to note the feedback and reciprocal relationships among virtually all of the items in the list.

- Effective planning, management, and efficiency
- Rational policy formulation and decision making
- The maintenance of human and aquatic health
- Environmental, ecological, and economic uses of water

- Rational allocations for municipal, industrial, and agricultural sectors
- Forecasting and managing floods
- Controlling pollution and contamination
- Scheduling of hydropower production
- Defining and apportioning the water of international rivers
- Resolving conflicts over international and transboundary rivers
- Adjudicating legal disputes over international and domestic waterways
- Cooperative and equitable use of domestic and international waters
- Research on long-term changes in the hydrological cycle
- Tracking demographic impacts on water and the environment
- Topographical, geological, and hydrographical research
- Determining the general hydrological conditions of river basins
- Determining cost ratios and good project and engineering designs
- Research on appropriate technologies and manpower training
- Tracking and managing forests and land use .

Data for these and other purposes, singly or in combination, are needed at one time or another on every stream in virtually every nation. They are required, as has been adumbrated, for informed, immediate decision making and for future planning and project designs [4, 5]. These needs can be satisfied only by having in place a good data gathering system and competent personnel to operate it and interpret the results. "For hydrologic science to move forward it is essential that data sets, once acquired, be properly identified and described...be catalogued and archived (including archive maintenance), and be made available to the scientific community at reasonable cost and effort." [5] Unfortunately, that capacity is all too frequently lacking in many parts of our interconnected world.

7. The Importance of Historical Water Data

In order to achieve a specified accuracy, some applications of data demand the development of records over a long period of time. Natural causes such as vegetation, sedimentation, and erosion, together with human interventions, change the characteristics of streamflow. Perhaps the most powerful forces that generate stream variations are weather and climate. These combinations of natural changes inherent in the flow of rivers along with human intrusions generate much uncertainty when the features of those flows are estimated. The longer the historical data record, the less the uncertainty [4].

There are other compelling reasons for long term data logs: long-term streamflow chronicles improve significantly the probability of recording extreme flood and drought events; these reports are of critical importance for devising measures to protect lives and property, and to mitigate the effects of drought [3].

Since a nation's socio-economic and agro-industrial advancement is dependent on its water supply, so will that nation's water infrastructure and data collection system mirror its upward growth. The progress of a nation can in many significant ways be tracked by the development of data on the flow, quality, uses, and consumption of its water supply. As water has become increasingly scarce and the environment more threatened in such places as the Middle East, it is imperative that accurate, long-term data are collected in such regions to determine how stream flow characteristics change over time due to such causes as agricultural practices, urbanization, groundwater development, or climate change.

8. Other Dimensions: Interdependency, Accuracy, and Scale

Worldwide technological and economic interdependence are growing every year. A function of this phenomenon is that advances in economic and social development, and in the spread of scientific and technological knowledge everywhere, are dependent more and more on systems of electronic and satellite data transmission and sharing.

Presently, even when water data are available, in many world areas owing to deficient collection technology and standards, or for political, security, or even cultural reasons, they must be treated with a robust suspicion. Until proven accurate, water data must be placed under the most rigorous critical scrutiny. Where water data are concerned, one is most likely to be working with percentages of accuracy, especially in regions such as the Middle East where water data are widely treated as classified information. Because water does not respect national boundaries, in the absence of international or basin-wide cooperation agreements, it becomes extremely difficult, if not impossible, to collect complete and precise data—especially in conditions of scarcity or maldistribution—for calculating accurate water balances and avoidance of conflict.

There is another, perhaps ancillary, facet of resource information involving certain perspectives of scale and time that deserves brief mention. The scale perspective of most people who work the land in poor, technologically under-developed nations is confined to the small acreage they are actually able to till. Although in relative terms this situation is improving, they are, with a few notable exceptions, unaware of national or global information that would provide them with the big picture and their relationship to it—information about health, weather, seeds, crops, insecticides, technology and how to use it, markets, money, family planning, communications, etc. Were they are supplied, as far as possible, with such information in real time and in terms accessible to them, they would be better able to decide on the most appropriate individual and collective behavior to improve both their immediate and long-term interests.

9. Key Factors in Data Gathering and Sharing

The collection and sharing of data have their own requisites if their potential is to be fully realized:

- Adoption of data collection standards
- Quality assurance for maintaining data collection standards
- Standardization of measurements, terms, and definitions
- Data comparability and consistency
- Maintenance of long-term (historical) data records

- Proper documentation, and collection of ancillary data
- Good processing, storage, and distribution (easy access and flow)
- Good staff training and water and air quality analysis services
- Research to improve data gathering technologies
- Data sharing requires:
    - data of known quality
    - data monitoring
    - data types and quality that are comparable over time
    - availability and accessibility of data
    - computerized databases with appropriate software to store, organize, and move data .

10. Basic Elements in Database Network Designs

Many of the problems associated with the gathering and distribution of water information will undoubtedly continue into the foreseeable future — particularly those that are rooted in political and security issues. But advances in data collection technology and communications science, and the creation of effective distributed information networks and pointer systems have already and will certainly persist in easing some of the difficulties. Distributed database systems in particular hold great potential for mitigating the problems of basin-wide and regional water management and distribution problems. To be most effective, data systems and network designs should incorporate certain key features. They should:

- Aim for an integrated distributed data system
- Take into account the important issues of
    - what are the most important specific objectives of the system
    - who controls and makes policy
    - who maintains the system
    - who benefits
    - what are the tangible user benefits
    - who pays
- Define the optimum data in relation to the attainment of objectives
- Establish the data requirements (i.e., users, needs, location of data, and where it is distributed)
- Set the specifications of the data (i.e., data standards, accuracy, and structures, the reporting format, the storing and accessing of data, and how the data is expressed)
- Create a system of data acquisition, quality control, and analysis
- Determine the best system of processing data and their products
- Put in place a good communications system, with appropriate, upgradable hardware and software components that connects the databases in the network and the system with a GIS (Geographical Information System) and other networks and has the capacity to collect and process a variety of spatial data in a variety of forms .

11. The MEWIN Model

In addition to the design components cited above, an effective distributed database network should also have certain basic features. Among the most important, aside from the elements already cited under previous headings, are:

- Simplicity in design and administration
- Careful, consensus-based planning and incremental development
- Layers of decision-making should be kept to a minimum
- Individual database managers in the network should maintain control of their databases and over what data would be shared
- Each member of the network should be enabled to recover any costs incurred by participating in the network
- Sophisticated network management, using relational software, which would:
  - coordinate among the satellite databases
  - respond quickly to a wide range of requests
  - harmonize and direct the flow of information among the network
  - create a "pointer system" to direct users to where data and information exists on the network .

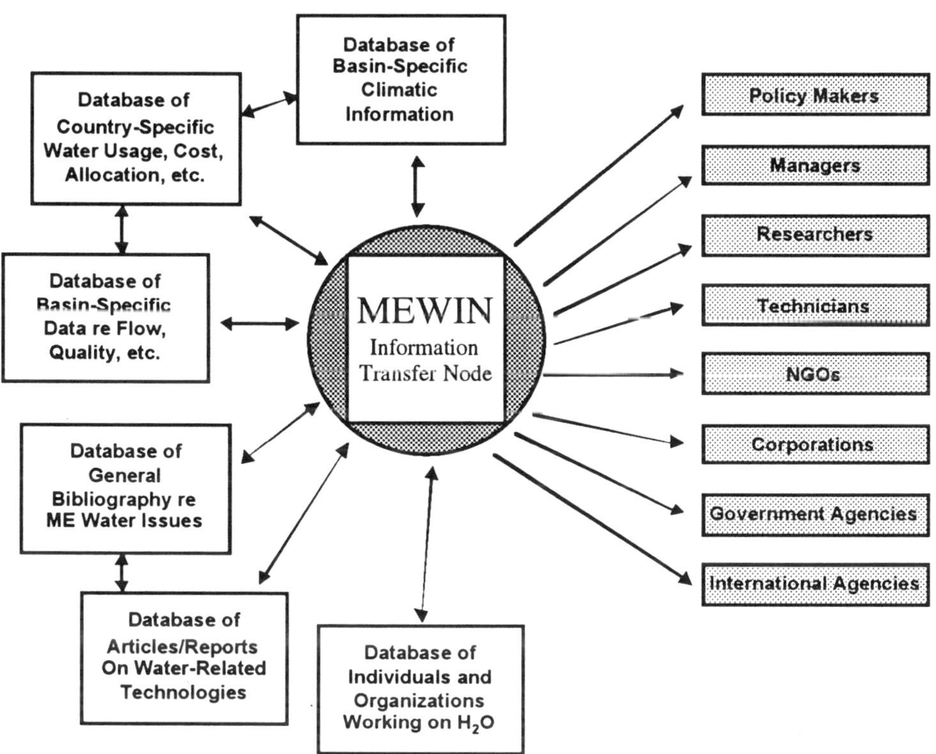

Figure 1. MEWIN (The Middle East Water and Information Network, University of Pennsylvania) has designed a model along these lines.

In this scheme there would be a MEWIN Information Transfer Node (MITN) that would act as the network manager. It would perform not only the coordinating, harmonizing, and pointing functions, but several others as well. For example, the MITN would:

- Standardize definitions and measurements and coordinate the establishment of common fields required by the relational software
- Prepare and distribute abstracts, indexes, lists, and reports, meld or process data and redistribute it throughout the network in electronic or various printed forms
- The Internet and GIS connections would allow MITN to update constantly, exchange, and share information throughout the network

and beyond. MITN could ensure that such activities would be carried out in purposeful, systematic, and productive ways.
- MITN could carry out the important task of implementing end-user rules, which would be drawn up collectively by the network members (these could be computerized and put on home pages over the Web).

Should such a network model succeed, it could create a synergy among its members, encouraging and enabling them to maintain and develop standards, adopt new technologies, enhance the contents of their individual databases, and introduce efficiencies into their operations. Each member could be informed of what others are doing and could be better able to keep up with new developments in the field of information sciences. The network would not interfere with the local operation or interests of the network databases; their data would remain intact under their own control while still functioning as part of the larger distributed database system. In time, the members of the network would become more closely connected with the mainstream of information and knowledge.

12. A Last Word on Overcoming Barriers to Data Sharing

There are many regions of the world where water-resource data is considered to be a security issue and is therefore classified to prevent its being shared. Sometimes the barrier to sharing is sheer bureaucratic inertia and inefficiency, and sometimes it is cultural attitudes that discourage sharing information because sharing could open a society to influences that are perceived to be corrupting. There are places where these three factors combine to raise the barriers even higher. The question therefore arises whether it makes sense in such circumstances, even to consider creating distributed database networks whose purpose is the sharing data and information. The answer is absolutely yes. There are a number of reasons to take a positive view:

- Economic and social development is driven by information
- All developing nations that seek financial and technical assistance must receive and give essential information to achieve their development goals
- Through assistance programs and other means, information inevitably reaches the outside where controls over access to it are eventually lost
- Computer and satellite technology together with the sciences of information and communications are steadily creating more sophisticated means of direct and indirect data gathering and data distribution through such systems as the GIS and the Web
- Water and other vital resources are such pervasive public and scientific issues that it is impossible to keep their data secret for very long. Water is involved in virtually all developmental projects and economic planning, in industrial and agricultural processes, in foreign and technological aid, and in energy
- Information and development are interdependent: there cannot be effective development without sufficient quality information, and

information is, in turn, an important by-product of development that makes continued development possible

- Much of the information and data related to water and the environment are not country or region specific. They appertain to basic and applied science, climate, precipitation, energy, technology, management, economics, demographics, transportation, law, etc.
- The great bulk of this information is in the public domain and can be pulled together with regional or national data that are available to create a huge pool of information in one easily accessible distributed database network.

In an interdependent world, no nation can isolate itself from the exchange of information that is integral to the communications revolution; if it tries, it will fail. Such a nation would be unable to move with the times and the consequence would be a progressive decline in its economic and social development. Ways must be found to persuade policy makers away from such dark and destructive tracks to the enlightened, self-serving path of cooperative, peaceful, and equitable resource management. And that will entail the sharing of data.

Notes

1. Simpson, M.R. & Oltmann, R.N. (1993) *Discharge-Measurment System Using an Acoustic-Doppler Current Profile With Applications to Large Rivers and Estuaries,* USGS Water-Supply Paper 2395, USGPO, Washington, D.C., 1 - 2.

2. Gleick, Peter, ed. (1993) *Water in Crisis,* Oxford, 117 - 119.

3. Paulson, Ricard (1996) Interviews and written communications, August to October.

4. Wahl, K.H., Thomas, O.S., & Hirsch, R.M. (1995) *The Stream Gaging Program of the U.S. Geological Survey,* USGS Circular 1123, Reston, VA 5 - 15.

5. Rantz, S.E. *et al* (1982) *Measurement and Computation of Streamflow,* USGS Water Supply Paper 2175, USGPO, Washington, D.C., 631 pp. This is the basic reference for stream-gauging.

6. National Research Council (1991) *Opportunities in the Hydrological Sciences,* National Academy Press, Washington, D.C., 214.

NOTES ON DATABASE MANAGEMENT SYSTEMS

J. S. MINAS
Office of Computing Services
Drexel University
Philadelphia, PA 19104 USA

Abstract

We are constantly reminded by the media of the vast amounts of information available to be collected, managed, analyzed, and applied to problems in all aspects of our lives. Along with the media coverage there is much talk about all the data we should access, manage, and utilize. However, there is not a commensurate understanding of the procedures involved or of the tools necessary and available for accomplishing these tasks. Managed data can be as simple as a record of household expenditures or as complex as a physical, chemical, and hydro-geological profile of a river basin. Regardless of the amount or the nature of the data, the principles underlying the process and the tools for accomplishing the task are the same. Before beginning any discussion of data management and analysis, it is necessary to understand the basic concepts of data management in order to establish a firm and stable foundation for drawing conclusions from the data.

Key Words

data, database, data bank, extraction, fields, formatting, management, records, reporting, SYLK

1. Introduction: Defining Data

It is surprising to discover that one of the shortest entries in the Oxford English Dictionary is the one for data. The OED entry states only that data is the plural of datum. The entry for datum is as follows:

> **datum** [L. datum given, that which is given]. 1. a. A thing given or granted; something known or assumed as fact, and made the basis of reasoning or calculation; an assumption or premise from which inferences are drawn. 1. b. Comb., as datum feature, level, line, -mark, -plane, -point, -year. 1. c. Philos. datum of consciousness, etc. Esp. datum of sense (cf. sense-datum). 2. pl. Facts, esp. numerical

T. Naff (ed.),
Data Sharing for International Water Resource Management: Eastern Europe, Russia and the CIS, 13–20.

facts, collected together for reference or information. 3. Use attrib. and in Comb. in the pl. form, as data bank, -handling, -transfer vbl. n., transmission; data capture; computing, the action or process of entering data into a computer, esp. when it occurs as an accompaniment to a related operation; cf. capture.

The second and third meanings are most widely used and the ones employed in this study.

Other relevant, associated terms are:

data bank

- a large collection of data stored for use or analysis, especially by a computer (*Oxford American Dictionary)*

- a fund of special information gathered and organized for quick use or analysis, usually on a computer *(Random House College Dictionary)*

- a fund of information on a particular subject or group of related subjects, usually stored in and used via a computer system (*Random House Dictionary of the English Language)*

database

- a large collection of organized information that may be updated and manipulated as needed *(New York Public Library Desk Reference)*

- a comprehensive collection of related data organized for convenient access, generally in a computer *(Random House Dictionary of the English Language)*

database management system (DBMS)

- a collection of hardware and software that organizes and provides access to a database. The computer program manages data, providing the mechanisms needed to create a computerized database file, to add data to the file, to alter data in the file, to organize data within the file, to search for data in the file, and so forth. *(Webster's New World Dictionary of Computer Terms*).

Another commonly used term is **data set**, which is virtually synonymous with **data bank** and **database**; the word **data** is usually understood to refer to these expressions.

2. Database Management Systems

A database management system (DBMS) is in its simplest terms a [computer] program for entering, editing, and managing the records in a database. Almost any general purpose computer may support some DBMS; microcomputers, workstations, and mainframes are all customarily used to support such systems. Several DBMS's are capable of running on many different platforms and operating systems. For example, ORACLE is supported on both microcomputers and mainframes; Claris's FileMaker Pro runs on both Macintosh OS and Windows platforms.

Creating a DBMS involves two steps: (a) defining the fields, and (b) entering the data. Defining the fields involves:

- Providing a ***name*** for each field
- Specifying the ***field type*** for each field
- Providing ***specifications*** for each field (optional).

Data are entered into named, i.e., defined fields. Entering the data may occur in a number of ways. Typically, the DBMS has a data entry screen — which is normally the same for each record — that shows a layout containing all the fields into which the data are typed. However, if the data already exist in some electronic form they may be imported by means of the import capability of the DBMS. Similarly, the records in one DBMS may usually be exported in some standard format for transfer to some other DBMS.

In order to simplify data entry and to reduce the likelihood of errors, specifications may be associated with a given field. Examples of such specifications are:

- Specifying a range within which the entered data must fall. For example, flow data may be constrained to a given date and/or between two specified point on the watercourse.

- Automatically inserting some particular date, say today's, for all records.

- Automatically inserting a serial number *n* indicating that this is the *nth* record created.

- Providing a list of 'legal' entries from which one must be chosen as the entry. This is especially useful in order to guarantee nominal uniformity of, say, the standards for measurement of water quality.

The way data are organized in a DBMS depends on the specific kinds of applications required and some of the inherent characteristics of the data themselves. Generally speaking, there are three logical database structures in common use. They are:

- The Hierarchical Database Model

- The Network Database Model
- The Relational Database Model.

In the **hierarchical model**, data are arranged in a top down multi-level structure in which any datum in a given level is associated with precisely one datum in the next higher level. Many traditional tables of organization are of this type. In such organizations each person has precisely one person to whom he or she reports, but there may be (and usually are) several persons in the organization who have several persons reporting to them. Such a structure is sometimes referred to as a *one-many structure*.

The **network model**, however, provides for the possibility of many-many relationships. For example, in most colleges each student takes several courses, and thus the relationship between students and instructors is an example of a *many-many structure*. However, in this example it is typically the case that the relationship between instructors and academic departments is a one-many relationship.

In both these cases, the relationships among the data may be represented in some graphical or tree-like manner. However, it is also possible in all such cases to give a 'flat' two-dimensional representation of these data. For example, one might construct a set of 'records' containing a student identifier (i.e., name or ID number), the name (or number) of the course, the name of the instructor, and the academic department (although the course identifier may uniquely identify the academic department). These data may then be arrayed in a record and field table as follows:

TABLE 1.

Student ID	**Course**	**Instr**	**Dept**
363-26-5690	M216	Smith	Math
363-26-5690	P316	Jones	Phil
189-36-7216	M216	Braun	Math
295-44-8725	H345	Grey	Hist
189-36-7216	P316	Jones	Phil
295-44-8725	P345	Black	Phil
276-44-8725	H345	Grey	Hist

Here, there are seven records and four fields. Each field has a particular type. In this example, the first field (student ID) is numerical, the remaining fields are date fields represented as *alphanumerics*. There are several other field types, including the possibility of graphic, video, and audio fields. Also, a field may be typed as a calculation field in which case its entries in a given record will be some mathematical or logical function of other fields in that record.

It is desirable, although not absolutely necessary for each field to be homogeneous; i.e., in a given field all the entries should be of the same type. If this is not the case, then some programming must be developed to render the entries homogeneous.

Table 1 gives information about students and the courses taken, but there are obviously other kinds of data that are relevant for the college's records. For example, the instructor' salaries, the students' addresses, etc. One way of making provision for these data is to augment Table 1 by adding additional fields to each record. However, continuing this process for the college's data needs results in a table of enormous size and one that would be used even in cases that require information from a relatively limited number of fields.

In order to deal efficiently with this problem, the **relational model** is frequently used. A relational DBMS arranges data into separate tables. In the above example, another table (e.g., Table 2) is constructed containing the students' addresses, and a third table could be used to give information about the instructors.

TABLE 2.

Student ID	**Street**	**City**	**State**
363-26-5690	12 Elm Street	Mytown	IN
189-36-7216	3276 Woodward	Yourtown	PA
295-44-8725	296 Highland St	Hertown	NJ
276-44-8725	1256 King Street	Histown	CA

TABLE 3.

Instr	**Phone**	**Salary**	**Ethnicity**
Smith	2-8953	$49,275	Asian
Jones	3-1617	$33,500	Hispanic
Brown	2-8936	$62,150	American Indian
Grey	4-7585	$45,000	White (non-Hispanic)
Black	3-1821	$28,500	Other

For some pairs of tables, there is a *common field* that provides the capability of linking the two tables and drawing upon the fields of both. For example, Tables 1 and 2 share the field of *Student ID*; Tables 1 and 3 have *Instructor Name* in common. These common fields enable the two tables to be linked relationally.

Each DBMS has some device by means of which the two tables to be linked may be identified. For example, in Claris FileMaker Pro the File Menu contains 'Define Relationships' as one of its options, the selection of which prompts the user for the tables to be used.

Whether it is preferable in any given case to use (a) one very large table or (b) a set of smaller tables linked relationally is a pragmatic question to be settled by a number of considerations, including the computer used, the speed required (often a hypothetical option *a* yields faster results than *c*), the ease of updating the table(s), and so on.

The examples used in Tables 1, 2 and 3, are, of course, only illustrative and are borrowed from a typical academic organization. However, the same considerations apply to virtually any situation in which the care and handling of data are of paramount importance. The nature of the fields defined in a DBMS will be determined by the nature of the context. For example, in the case of the management of water resources, some typical fields, among others, with typical values, might be the following for, say, the Jordan River basin:

Basin Location: Jordan, Israel, Palestine, Syria

Precipitation: long-term average measured in millimeters per year

Streamflow: adjusted average as measured in million cubic meters per year

Ground Water (renewable): average number of million cubic meters per year and average rate of renewal

Desalinated Water: number of million cubic meters per year and cost per cubic meter.

3. Extracting and Reporting

One of the main reasons database management systems are created is to provide an efficient and standardized means in which data may be extracted in an organized and informative way. Reports and derived data sets are the primary vehicles for this function. The difference between the two is that a **derived data set** is simply a file containing a specified set of records and fields and a **report** consists of a derived data set plus some special formatting, headers, column totals, averages, etc. The procedure for creating either a derived data set or a report is basically the same and involves the following steps:

- Specifying the subset of fields to be included
- Specifying the subset of records to be included
- Specifying the sorting sequence required
- In the case of reports, specifying the special formatting required

All these specifications may be "canned" (stored, filled) in one way or another and given a name so that the data set or report may be called up at any time. One of the major advantages of a DBMS is that a particular extract or report may be created repeatedly in the same form but with different contents as the data are updated or revised.

In addition to creating reports which represent records and fields consistent in format and accurate as data updates require, reports can also be improved in many software products enabling the addition of graphics, logos and text to explain and expand on the basic information. Software products also enable the easy integration of reports into word-processed documents such as annual reports and research reports. Electronic documents such as these are dynamic and as the data in the original data reports are updated, or changed, those changes will also be reflected in the larger document as well.

What has been described thus far is primarily the method of organizing and presenting raw data. However, the real utility of databases is the ability to analyze the data to answer questions, draw conclusions, or make hypotheses. Exported database information can be analyzed by well-established programs such as SPSS, SAS, and Oracle to determine frequencies, patterns of occurrence, relationships between and among occurrences and comparisons of historical information. The ability to analyze data and draw conclusions from the analysis becomes the basis for decision making on all levels, from personal to those related to medical research and governmental policy-making.

4. Data Export and Import

Frequently, it is useful (or necessary) to move data from one DBMS (S1) to another (S2). This task is accomplished by means of exporting and importing. A new file (F) is created by exporting the data from S1 using some particular file format for F that is supported by S1. It is also necessary that the particular file format is one used that is also supported by S2. Then, the file F may be imported into S2. This procedure may be used in moving data around from-and-to a DBMS and to-and-from other types of applications such as spread sheets. In all such cases, the choice of the file format used is important; there are several more or less standard file formats for importing and exporting.
Typical examples of such file formats are:

Tab-Separated Text files have a tab character inserted between each field of each record and a return character between each record

Comma-Separated Text files are similar to tab-separated text files except that commas are used instead of tab characters and all fields (except unformatted numerical fields) are enclosed in quotation marks

SYLK (Symbolic Link Format) formats originated with spreadsheet data primarily for numerical data; the data are arranged in rows and columns, each row representing a record and each column a field.

There are several other import-export file formats that are used in a more or less proprietary way. For example, **DIF** is an earlier spreadsheet format, **WKS** is a special Lotus 1-2-3 format, and **DBF** is used primarily with dBase III.

Moving an exported file, F, from S1 to S2 may be accomplished in a number of ways. Traditionally, the device of shipping some physical medium, containing F, such as a disk or tape, from S1 to S2 has been used. Implementing the transfer electronically has been, until recently, limited to cases in which S1 and S2 reside on the same computer, or the same local network, or some proprietary network (e.g., BitNet). However, the Internet, together with a number of Internet-based utilities such as *fetch* and *sendfile* make such transfers relatively easy. The Internet is rapidly becoming an international and universal resource. However, the security problem remains a concern, especially with highly proprietary and sensitive data. Rapid and effective advances in security protection are being made, and the problem should soon be eliminated.

5. Conclusion

Record keeping and information management have historically been an important part of human development and progress. The ability to manage data accurately and easily in electronic databases has enabled the study of information by anyone — not just computer scientists or research specialists — in ways never before possible.

Understanding database structures and analysis is important for decision and policy makers and for managers. Comprehension of these issues promotes a deeper understanding and effective use of information.

6. References

Cannon & Meyer, 1997. *Building a Better Data Warehouse.* Prentice Hall Press.

Connolly, T.M., 1995. *Database Systems: A Practical Approach to Design.* Addison-Wesley Publishing Company.

Fortier, P.J. (editor), 1996. *Database Systems Handbook.* McGraw Hill Text.

Gorman, M.M., 1993. *Database Management Systems.* John Wiley & Sons.

Harrington, J.L., 1997. *Database Management for Microcomputers.* Holt, Rinehart & Winston.

Heller & Stonebraker, 1998. *Readings in Database Systems.* Morgan Kaufman Publishers.

Hernandez, M.J., 1997. *Database Design for Mere Mortals.* Addison-Wesley Publishing Company.

Larson, J.A., 1995. *Database Directions: From Relational to Distributed.* Prentice Hall Press_

Teorey, T.J., 1994. *Database Modeling & Design.* Morgan Kaufman Publishers.

Zaniolo et al, 1997. *Advanced Database Systems.* Morgan Kaufman Publishers.

HYDROLOGIC DATA COLLECTION FOR RIVER BASIN MANAGEMENT

RICHARD W. PAULSON
U.S. National Weather Service
1325 East-West Highway
Silver Spring, Maryland, 20910 USA

Abstract

There are many different types of hydrologic measurements — largely of water quality — that are made in support of river-basin management. The maintenance of data collection programs of these measurements is important. Even the collection of data on a small number of relatively easy measurements can provide valuable water resources information to river basin managers. Basic to the collection of these data are the adoption of data collection standards by the collecting organization, operation of quality assurance programs to affirm that the data are collected properly, and support programs to train staff, provide water quality analytical services, instrumentation support, and research to improve data collection. There are many hydrologic services that successfully operate historical and real-time data collection programs whose data and publications are becoming available on the global computer network known as the Internet.

Key Words

ADCP, archiving, AVM, comparability, DCP, data, discharge, GIS, GOES, HOMS, hydrology, management, measurements, quality, remote sensors, rivers, technology

1. Introduction

Data are the basis of information. Hydrologic data are the basis of most of the water-resources information that managers need if they are to manage river basins and meet demand for municipal, agricultural, and industrial water use, and provide water for hydropower generation and aquatic habitat. Satisfying these demands, and protecting and conserving water resources is complicated by several factors: variability in the precipitation delivered to a basin, changes in water use and land use, ground-water mining, long-term climate change, and other natural and human induced changes in the spatial distribution, variability, and quality of water. In well developed river basins, the operation of dams, reservoirs, canals, irrigation projects, hydropower generating

T. Naff (ed.),
Data Sharing for International Water Resource Management: Eastern Europe, Russia and the CIS, 21–44.

facilities and other engineering works also affects the distribution, variability and quality of water as it makes its way to a basin's outlet. In lesser-developed basins, human populations are more vulnerable to the uncontrolled extremes of flood and drought.

There are scientific challenges to the collection of data on the distribution, variability, and quality of water in a river basin, to the development of models or other analytical techniques for describing basin hydrology, and to providing a rational basis for decision making by river basin managers. These challenges are more difficult if the basin is international and the data collected in the basin are not comparable. This, commonly, is the case if several countries collect data in their own political jurisdictions and use different data collection standards, instruments, and analytical techniques, and store the data in different types of databases. This problem is compounded if the countries of a shared basin do not maintain records of their standards and techniques. Such records provide a basis for comparing data records from different countries and the information derived from the data. Problems of comparability are minimized if organizations agree to collect the data in conformity with the same standard, use instruments and analytical techniques that meet the standard, store data in data-bases that are accessible by others, and maintain good records on their data collection techniques.

Hydrologic data must not be discussed in isolation from related data and issues. They must be considered in relation to meeting specific water resources information needs and within the context of the technologies and resources an organization requires to collect the data. This essay will employ that approach in describing such associated matters as water resources information needs of river basin management, hydrologic data collection and analysis that provide the information to meet those needs, the technologies that are used to collect the data, data collection standards, training, quality assurance, logistical issues that must be addressed by hydrological services, and sources of information about hydrologic data collection, all with a view to clarifying these issues for planners, economists, managers, and others who need to know about water resources, but are not necessarily scientists or engineers.

Moreover, such readers need to understand the value of hydrologic data and the resources that hydrologic services require to sustain hydrologic data collection programs. Without such resources, data collection programs collapse or produce data that have questionable value. This study is not an exhaustive treatise on hydrologic data collection, rather it is a summary of the most common hydrologic data that are collected (citing selected data collection examples) and a description of the technologies and analytical techniques that are coming into use around the world, including weather radar and satellite estimation of precipitation, and Geographic Information Systems (GIS). A major purpose of this study is to advocate uniform, shared data-collection standards, support programs that assure quality data, staff training, the provision of laboratory and instrumentation support, and the application of research new methods.

2. Water Resources Information

If one were to measure the volume rate of water flowing down a river and record that discharge once an hour, then the hourly discharges would be data. If one were to continue to measure and record the data every hour for many years, it then would be possible to analyze the discharge data record and characterize the flow characteristics of the river. The characteristics could include the probability of occurrence of a drought or a flood of a particular magnitude, the average monthly discharge for each month of the year, and so forth. The flood and drought frequencies and the average monthly discharges are information. That is an example to how data serves as the basis of information.

If one also were able to collect data and analyze them at many locations on rivers in a basin, one could characterize how the basin operates as a water-discharge system. This provides useful information for planning and operating basin-wide water supply and flood-control measures. If the data and the derived information are of poor quality, then basin managers will be less effective in making decisions based upon them. Hence care is needed in data collection if the derived information are to be trusted and have value.

Data are approximations of reality. No matter how carefully they are made, river discharges measurements are always taken with the realization that the river discharge data will fall within a band of uncertainty around the actual discharge. Similarly, water level data, water-quality laboratory analyses, and other water measures are approximations of reality too. There is a body of science that studies how well data approximate reality and one influential factor that must be taken into account is the tradeoff of cost versus uncertainty when setting up data collection programs.

Data also are collected by fallible human beings, and by instruments and techniques that have limitations. Care must be taken to minimize the effects of errors in human judgment and performance, and degradation of instruments and techniques so that they do not further increase the uncertainty of these approximations of reality. Quality assurance programs are essential and intended to maintain data collection programs within acceptable uncertainty levels.

Because data are the basis of information, data-collection programs must be established with clearly defined information needs as illustrated in Table 1.

TABLE 1. Typology of water information needed for river basin management.

Need Category	Purpose
Aquifer Water Levels	Define aquifer capacities to provide the base flow of river during dry periods and residential self supply in many rural areas.
Snow Pack	Define the amount of water immobilized as snow and ice, which is a principle source of water for use in spring and summer in many basins.
Soil Moisture	Define the amount of water in the soil, which affects the amount of water that runs off during a rainstorm as well as agricultural potential.
Water Distribution	Understand the quantity of water in streams, rivers, estuaries, surface-water bodies, and aquifers in a basin, both historically and at the moment.
Water Movement	Define the rate at which water is flowing from place to place.
Water Quality	Define the biological, radiological, and chemical characteristics of a water source.
Weather	Forecast temperatures and precipitation, which assists managers to anticipate changes in water demand and to control floods.

Maps, of course, are also among the more important informational needs as Table 2 demonstrates.

TABLE 2. Typology of map information needed for river basin management.

Need Category	Purpose
Boundaries	Define property lines and city, county, state and national boundaries, which govern the rights to water and sets political jurisdictions.
Flood Forecast Points	Show potential flood-reference points during times of extreme river flows.
Flood-Prone Areas	Show areas that are prone to flash floods or are in a flood plain, and identify the population and facilities at risk during floods.
Geology	Shows the areal distribution, thickness and age of principal ground-water aquifers and how they affect the quality and availability of ground water.
Hydrologic Stations	Show the locations of precipitation, river gauging, water-quality, and ground-water stations, and other sources of hydrometeorological data, which are sources of historical and real-time data.
Industrial Sites	Identify actual or potential sources of hazardous waste and contamination and points of industrial water withdrawal.
Land Uses	Identify how land is used, either in a natural state or in human affected states of urban, agricultural, and other development which affect the quantity and quality of surface water runoff or ground water.
Population	Population density and distributions show where people live, and points of municipal water withdrawals.
River Networks	Show the interconnections between surface-water features, such as rivers, water bodies and wetlands, which define paths that surface waters follow.
River-Basin Boundaries	Boundaries define the geographic limits of the natural flow regimes of the rivers. [1]
Soils	Quantify how soils influence rainfall-runoff relationships and the quality of the water that permeates the soil
Special Features	Show the locations of national parks, water shed of municipal water supplies, wildlife sanctuaries, and other areas of special consideration.
Topography	Defines the slope of the land, which affects how quickly water will run off after precipitation occurs.
Water-Control Structures	Show locations of dams, reservoirs, canals and flood-control levees, which are among the tools that are available to redistribute water in space or time.

Managers also require reports that interpret data and add value to the knowledge of basin hydrology. A sampling of such reports are shown in Table 3.

TABLE 3. Typology of Interpretive Information Needed For River Basin Management

Content	Purpose
Causal Relationships	Document relationships between land use and water quality or between land use and flood frequency and estimate the effects of human development on water quality and runoff.
Climate	Characterizes the annual and long-term patterns of water availability from precipitation.
Drought	Estimates the severity, probability of occurrence and geographical distribution of low river discharge and droughts of particular magnitudes and duration.
Flood Frequencies	Estimate the severity and probability of occurrence of high river levels and large discharges.
Groundwater Assessment	Estimate a sustainable yield of groundwater withdrawals.
River Routing	Estimate the time it takes to route flood discharge down through the drainage network or convey a pollution spill to travel throughout a basin.
Standards	Define water quantity, water quality and other legal requirements to maintain water resources within specified limits.

Many information needs are not readily defined by existing maps and reports, and water-resources scientists and managers may need to conduct *ad hoc* analyses of data to solve unique problems. Hence they need to access databases where hydrologic data are stored. Rarely are all of the data that are the foundation for the information listed above available from one organization in a basin. Hence, interagency coordinating mechanisms, earth-sciences libraries, data-sharing agreements, and other data sharing mechanisms are important ways to provide the needed information to river-basin managers. An excellent example of data sharing is the California Data Exchange, where numerous local, state, and federal agencies share on-line hydrologic data, radar imagery, scheduled reservoir releases, snowpack information and other conditions that are critical to managing river basins in California. [1]

3. Hydrologic Data Collection

There are three basic steps in hydrologic data collection: make the hydrologic measurement itself; record the measurement; and convey the measurement record to a data repository.

In regard to the latter step, there may be a large interval between the time of data collection and arrival at a data repository, or it may be conveyed almost instantaneously.[3] Except for the spatial data that are represented by the maps cited above, most of the information cited in Section 1 above are derived from the analysis of a relatively small number of types of hydrologic data, such as water level, river discharge, basic water-quality measures, precipitation, and water use. Unfortunately, vandalism of data-collection equipment is a problem almost everywhere and it must be taken into consideration in designing water- monitoring systems.

3.1. WATER LEVEL

By far, the most widespread quantitative measure of groundwater and surface water resources is water level. It is measured as the depth of water with regard to an elevation reference, such as meters above or below an elevation reference point.[4] In the case of ground water, observation wells that intersect important aquifers are used to measure the depth to water in the aquifer from a reference level selected as the elevation of the land at the well. If such measurements are taken and recorded over time, one can determine how the water level in the aquifer varies in response to water withdrawals, seasonal fluctuations in natural recharge, and — if the record is long enough and the area is not disturbed locally by human activity — climate change. In the case of surface water, water-level measurement in lakes and reservoirs can be used to monitor the volume of water in these water bodies. In a river, water level can be correlated with river discharge because high water levels imply high discharge. However, the actual relationship between water level and discharge can be quite complex and the subject of much analysis.

3.1.1. *Water-Level Technologies*

The oldest technology for measuring water level is the human observer, a reliable technology that is still in widespread use in many parts of the world. Even in countries that use more sophisticated technologies, there is still reliance on observers who make manual water-level measurements from time to time to assure the quality of automated water-level monitoring equipment. Commonly, a large, precisely graduated ruler, known as a staff gauge is permanently secured in a hydrologically strategic place where an observer reads and records the water level manually. Alternatively, an observer may make a measurement to the water surface from a bridge by carefully lowering a weight until it touches the water surface. In either case, the observer manually reads the water level, records the water level in a book or on a form, and conveys the data record to a central repository.

In automated systems, water level is most commonly measured by tracking a float that moves up and down on a water surface. [1] Placing a float on the surface of a flowing river is problematical because the water is moving, there may be waves on the water, debris may be moving down the river, the river may be ice choked, and so forth. Consequently, hydrological services use stilling wells, which are shallow wells constructed adjacent to the river. The wells are connected to the river by one or more intake pipes and the water level in the well is equivalent to the water level in the river, but without most of the problems of measuring open water. Stilling wells present their own problems because over time they may become choked with sediment or they may freeze in the winter. Hence, stilling-well maintenance is an important activity in river measurement.

In a stilling well, the distance from a fixed shaft to the float is monitored by a wire that is tethered to the float and runs over a shaft to a counterweight. Once the system is calibrated, shaft position becomes a measure of water level. The shaft may be mechanically connected to a pen that makes a trace on a paper that moves slowly at a fixed speed past the point of the pen. The pen draws a line that represents the change of water level over time. Such a shaft-encoding system includes the measurement of water level (the float and shaft) and an analog recording of the measurement (the ink-and-paper record).

Other water-level measuring technologies are also in widespread use, such as pressure transducers, manometers, and acoustic rangers, which may measure water level in a stilling well or in the river itself. Pressure transducers are electronic devices that produce electronic signals that are related to the pressure on the transducer. In the case of hydrology, the transducer is fixed well below the water surface and the pressure on the transducer is related to the weight — and therefore the height — of the water above the transducer. The transducer's signals are monitored and recorded as the water levels vary over time.

In the case of a manometer, the open end of a tube is placed below the water surface and nitrogen gas is slowly bubbled from the tube into the water. The back pressure on the tube is monitored in an adjacent shelter with a pressure transducer or other type of system that balances the pressure electromechanically.

An acoustic ranger is a device that periodically sends a sound pulse to the water surface and measures the time until the reflected sound pulse is sensed at the ranger. The elapsed time is a measure of the distance to the water surface. In the case of the pressure-based measurement systems, the systems must compensate for changes in barometric pressure and in the case of the acoustic ranger, the system must compensate for air temperature, which affects the speed of sound through the atmosphere. With these three technologies, the measurement may be made in a stilling well or in the environment.

In recent years, the recording medium for shaft encoders, pressure transducers, manometers, and acoustic rangers has become electronic memory devices known as solid-state data loggers. The loggers are replacing punched paper tape recorders or analog strip chart recorders that were in widespread use for many decades. The advantage of punched paper tape or electronic recording of the data is that the data can be read automatically and transferred to the data repository more efficiently than manual readings or manual interpretation of ink-and-paper analog records. The latter also are more subject to human error.

There are many factors to consider in selecting water-level measurement and recording technologies, including the environment in which the measurements must be taken and the cost, complexity, reliability, precision, and staff skill requirements.[4]

3.2. RIVER DISCHARGE

Measurement of river discharge is very important for flood forecasting, water-supply, hydropower generation, and river management in general. It is measured as the volume of water that a river carries past a point during a given time, such as cubic meters per second. It is one of the most difficult and expensive of hydrologic measurements because the pattern of water velocity is highly variable throughout the cross section of the river at any discharge site and must be accounted for.

3.2.1. *River-Discharge Technologies*

The traditional way to measure river discharge is manually, by making a series of water-velocity measurements with current meters. [1] Using this approach, water velocity measurements are made at selected depths in numerous sections throughout the river's cross section. The measurements may be made by a streamgauger who wades across the river, stopping at predetermined points and making water velocity measurements at two or more depths.

For larger rivers that are too deep to wade, the measurements may be made from a boat, off a bridge that spans the river or from a small platform on a cableway that is constructed specifically for discharge measurement. (Unfortunately, river discharge measurements can put the hydrologist's life at risk because the debris that rivers convey during floods can entangle current meters placed in the water and literally drag cable cars into the water.) In all cases, the velocity measurements in a section are used to calculate the discharge in each section of the river. The discharges in all the sections are then added up to estimate the total river discharge.

At many river monitoring sites, it is not practical for staff to be present at all times and river water level is monitored as a surrogate of river discharge. If the geometry of the river channel is stable, hydrologists make manual measurements of discharge at numerous discharge levels and relate the discharges to the respective water levels. [3] Such a river level-discharge relationship is used to estimate river discharge when the water-level data are processed at the data repository. At many river monitoring sites the river geometry is stable and infrequent river discharge measurements are required to assure that the river level-discharge relationship has not changed.

At other sites — chiefly after floods have mobilized and deposited sediments — the channel geometry changes enough to invalidate the pre-flood relationship between river level and discharge, requiring the hydrologic service to make more frequent measurements to reestablish the river level-discharge relationship. Unfortunately, at many river sites the river channel is largely comprised of mobile sediments making the channel geometry unstable. In those instances, the river level-discharge relationship also is unstable.

Where stable river level-discharge relationships are not possible, hydrologic services may construct hydraulic structures in the river itself, forcing the river to flow over or through a structure with known hydraulic characteristics. In these cases, water level or some other physical characteristic of the water can be monitored and the river discharge computed, based on the hydraulic characteristics of the structure. Constructing such structures can be a significant capital cost, especially if the river is large. In a more extreme measure of this approach, operators of dams can measure the flow of water through hydraulic structures in the dam or a hydroelectric powerhouse and compute the discharge of water to the downstream outlet of a reservoir.

There are acoustic technologies in use that automatically measure river discharge directly. This Acoustic Velocity Meter (AVM) technology uses sensors — i.e., transducers — that transmit and receive acoustic pulses in water. [4] If a transducer on one side of a river at a mid depth transmits a pulse to another transducer on the other side of the river, and the second transducer is upstream or downstream from the first, the time it takes to transmit a sound pulse in one direction will be different from the time required to transmit a pulse in the other direction. The movement of the water in the river causes this time difference and the length of the time difference is a measure of the average water velocity across the channel direction.

Those acoustic velocity meters measure average velocity at a fixed depth. If the river is deep and there are significant vertical river velocity gradients, two or more sets of transducers are installed at different depths and the river discharge is estimated from two or more time differentials. At locations where acoustic systems are in use, traditional river-discharge measurements are made from time to time to establish the relationship between discharge and acoustic system measurements. The capital investment in acoustic velocity meters is substantial, but provide automatic and real-time estimates of river discharge.

In recent years, Acoustic Doppler Current Profiler (ADCP) systems have been developed that measure the vertical variation in horizontal velocity in a river from a point on the surface to the river bottom. Transducers on these boat-mounted systems send four narrow-beamed pulses of acoustic energy from the bottom of the boat down into the water column, then detect reverberations of sound from sediments and other objects being carried by the moving water below. The reverberations return to the transducer at a slightly different frequency that represents a Doppler shift induced by horizontal movement of the water. By recording lag time and the reverberations, the system estimates the horizontal velocity of the water at numerous levels beneath the boat. If the boat has an accurate way to track its position, the boat can traverse the river, record the acoustic data, and automatically compute river discharge at the end of the traverse.[5] Although these boat-mounted systems are expensive, they provide a quick, reliable, and relatively safe way to measure river discharge.

Limits on the use of acoustic velocity or discharge measuring systems include cost, complexity, the need for highly trained and experienced staff, and competent private sector companies to supply and maintain the systems. Additionally, acoustic systems may not be applicable if sediment concentrations are excessive because suspended sediment particles absorb acoustic energy. Both types of systems allow acoustic data to be recorded by data loggers for subsequent review and analysis.

Electromagnetic technologies have been used to measure water velocity for many years. Water is a conductor and, as it moves through the Earth's gravitational field, it will generate an electric potential at right angles to the direction of water flow and the magnetic field. However, the electric potential is very weak and there is much electrical "noise" that renders such a passive technique useless. Nonetheless, the principal is valid and commercial water velocity meters that provide their own strong magnetic fields have been used for years to measure water velocities in closed pipes.[6] The USGS river gauging program has produced considerable data which has been used extensively for flood and drought estimation. [8]

3.3. WATER QUALITY

Thousands of water quality measurements are collected annually by hydrologic services, but only a few can be discussed here. These measurements include the concentrations of sediment particles that are conveyed by the river in suspension or along the river bed, concentrations of dissolved chemical or radiological constituents of substances in the water, biological conditions of water, and the state of the water itself, such as water temperature and pH. Many of these measurements are made in order to compute the quantity of contaminants that a river conveys—so-called constituent loads; in order to compute such loads, river discharges must be collected concurrently.

Moreover, water-quality contaminants commonly are attached to sediment particles requiring that data-collection programs in support of computing contaminant loads measure the contaminant concentration in the water as well as contaminants residing in the sediment that the river is conveying. To make these programs even more challenging, most of the sediment that moves through a river system is conveyed during relatively brief periods of peak discharges.

Rivers are rarely homogeneous; therefore, vertical and horizontal variations in water quality as well as water velocity are to be expected. Such variability results from several causes such as the effects on water quality from tributary flow, varying concentrations of suspended sediment that exist from the river bottom to the surface, and the effect of sunlight on biological activity.

Furthermore, in lakes, reservoirs, and other standing bodies of water, stratification of water quality is to be expected during much of the year. Hence, care must be taken to devise a data collection strategy that considers (1) the variation of water quality in the water body itself, (2) the location of the water-quality sampling location with respect to sources of contamination, (3) the temporal variation of water quality due to the daily cycle and seasons, and (4) the effects of river discharge on water quality.

3.3.1. *Water Measurement Technologies*

Most water quality data collection has a manual component. There are a relatively few measurements of water quality that can be made automatically. The most common automated measures of water quality are dissolved oxygen concentration, temperature, pH, and dissolved solids concentration (See, for example, http://www.campbellsci.com/sensors.htm). Because of the spatial variability of water in a water body, these measures can only be viewed as indices of water quality. Despite these limitations, such measures are highly useful to gain insight into the time variability of water quality during diurnal and seasonal cycles, during extreme events of floods and droughts, or to alert authorities to unusual river conditions.

As with the measurement of water level discussed earlier, it is difficult to place automated sensors of water quality in a river itself; consequently, water may need to be pumped from the river to bathe the sensors in a protected chamber. At water quality stations where data are collected automatically, data are recorded on site by means of a variety of paper-based or electronic measurements and conveyed to data repositories manually or automatically.

Most commonly, water quality is determined by measuring the characteristics of water samples that are collected manually and analyzed by the side of the river or at a water quality laboratory. In the latter case, care is required to preserve the water sample and convey it promptly to the laboratory.

Staff must be highly trained and properly equipped to sample water bodies. Where there are vertical (depth) variations in a water body, hydrologic services commonly use depth integrated water quality samplers that sample continuously from the bottom to the surface of the water, or samples are taken at selected depths in the water. In the latter case, the water samples may be added to a composite container. Where there are horizontal (topographical, surface) variations in a water body, hydrologic services sample at numerous points across the body of water. Again, water samples may be composited or kept separate for analysis.

Many agricultural and industrial chemicals, metals, and other toxic chemicals are dangerous to human and aquatic life in even the most dilute concentrations. Great care must taken in sampling water bodies for these constituents. Care also must be exercised in protecting the water samples from contamination in the field, during transport to a laboratory, or in the laboratory itself.

There may be numerous analytical techniques available in a laboratory to measure each characteristic of water quality. Since each technique approximates the actual value of a constituent concentration within a known range, care must be made in selecting and recording the analytical techniques that are used. There are numerous examples where an apparent trend in a characteristic of river water quality over time was detected by mathematically analyzing data from a water quality laboratory that adopted a new analytical technique during the period of record. In such instances, because the introduction of a new technique created a step function in the recorded data, the mathematical technique detected a trend in water quality in the river itself. Water-quality laboratories from numerous organizations form associations and participate in quality assurance procedures — such as conducting comparative analyses of common samples — to assure users of the laboratories that their analytical techniques and staff work within documented ranges and precision.[7]

3.4. PRECIPITATION

Precipitation is the source of renewable water resources in a river basin and is measured in the depth of water that falls on the land surface during a period of time, such as millimeters per day. Precipitation may fall as rain upon the basin or it may fall as snow or ice. In either case, precipitation is an important constituent because it contributes immediately to water availability in the basin if it is rain or to subsequent water availability if it is snow or ice. Precipitation may occur from large cyclonic storms that tend to bathe large areas in similar quantities of precipitation or convective storms that tend to bathe relatively small areas with intense precipitation. In the latter case, the precipitation may not even be recorded if the convective storm misses data-collection sites.

Measurements are made of precipitation by collecting and accounting for the water at a particular point or by making assessments of the mean areal precipitation of an area. In the former case, manually read accumulations or an automated measure collected at a point provides a sample of the precipitation that falls. In the latter case, measurements are made of precipitation by analyzing radar data or satellite imagery. Rarely should areal measurements of precipitation be considered valid without benefit of point measurements that are used to calibrate areal estimates.

Measurement points must be selected carefully because precipitation can be highly variable in time and space. The points must be selected so that precipitation is not collected near buildings, trees, and other sources of interference. Moreover, elevation is an important consideration in measuring precipitation because reductions in pressures and temperatures of air masses that are carried to higher elevations induce water saturation and precipitation. In the extreme, temperatures may fall below the freezing point and the precipitation may fall as snow or ice. Hence, stations must be selected carefully to take into account so-called orographic effects on precipitation.

3.4.1. *Precipitation Technology*

In its simplest form, precipitation technology may consist of no more than a container that is used at a precipitation station to collect water that falls as rain whose quantity in the container is recorded from time to time. Where data are collected manually, staff examine the container periodically and note the amount of rain that has accumulated since the last reading. In an automated station, the level of the accumulated water in the container is monitored, as discussed in the section on water level above. Additionally, some rain gauges may measure the weight of the accumulated water.

Measuring the level or weight of water is problematical because some of the accumulated water may evaporate. A common technology to counter this problem is the so-called tipping-bucket rain gauge. This is a device that contains two small adjacent containers, such that the rain falling upon the gauge is directed to one of the containers that fills with water. Once it fills after a predetermined amount of rain — say 1 millimeter — gravity causes it to tip and empty, and automatically position the second container to accumulate the subsequent rain. The raingauge has an internal clock that records the times that the buckets tip. Thus, the gauge records the absolute time and interval during which a known quantity of rain fell. This is especially useful during periods of intense rain when good definition of the intensity and time distribution of rain is important to record.

Where the precipitation is frozen, hydrologic services attempt to measure the water content of the accumulated snow and ice. In many semi-arid locations of the world, snowmelt is the principal source of water for human use and it is important to know the potential runoff before the spring agricultural growing season begins. Where the data are collected manually, staff may make periodic visits to predetermined sites throughout the winter and spring, and sample and weigh the snow along a cross section of the snow pack at the site. These staff are at risk because of the need to sample where the snow accumulations may be substantial, temperatures are low, the locations remote, and winter weather conditions variable.

Because of the risk and the need to know the conditions at such sites more frequently than is possible from manual measurements, some hydrologic services may use frequent areal or satellite photography of snow accumulation areas, or operate automated snow monitoring sites. It is difficult to monitor snow and ice automatically. One approach is to melt the snow and turn it into liquid water automatically, and measure it as one would measure rain. At a remote automated measuring site this requires a substantial energy source to melt the snow and antifreeze to keep it from freezing again.

Another approach is to measure the accumulated weight of the snow with a snow pillow. This is a flat container that is placed on the ground at the snow-measurement site before the onset of the snow season. The snow pillow contains a liquid that will not freeze and will rise up in a tube as the snow accumulates and compresses the pillow. One then estimates the weight (and the water content) of the snow above the pillow by measuring the elevation of the liquid in the tube. These techniques are valid in principal, but difficult to use in practice because snow is not homogenous, and alternating strata of snow and ice may actually form structures over the pillows and degrade the quality of the measurements

Other approaches to measuring snow are (1) to position a radioactive isotope at the land surface and measure the attenuation of radiation by the snow from a site well above the top of the snow surface or (2) to place an acoustic sensor well above the top of the snow surface and use acoustic signals to measure the distance to the top of the surface.

Regardless of the automatic measurements of snow and ice that are made at such sites, the sites invariably collect other types of meteorological data, such as air temperature, barometric pressure, solar radiation, wind direction and speed, relative humidity, and other parameters. There are many well-established technologies to collect these meteorological data which they are not listed here. There are important hydrologic implications for collecting these data along with water content of snow. Weather conditions on the snowpack affects the rate at which the snow may be evaporating or melting and releasing water for runoff. Among the most important contributions to flooding is warm rain falling on snow because the rain releases water in storage as snow and ice, and, at the same time, delivers new water to the basin.

The chemical quality of precipitation has become an important issue in many areas that are affected by sources of atmospheric contamination. In many of these areas, the acidity and concentrations of chemical constituents in rainwater are so serious that they adversely affect the quality of water in river basins and the suitability of the water for maintaining aquatic habitats and human use. This is a topic that is beyond the scope of this paper, but manual and automated sampling and analysis programs are common in North America and Europe.[8]

Weather radar systems have come into use in recent years that permit the tracking of storms and making areal precipitation estimates. The most advanced of these radars are so-called Doppler radars which collect radar reflectance data to determine areas of atmospheric convergence and divergence and estimate the moisture content of the atmosphere. Areas of convergence and divergence are especially important in monitoring and predicting the occurrence of severe weather and convective precipitation. Areal estimation of precipitation via weather radar is difficult and must be normalized by the inclusion of so-called ground truth precipitation data from reporting rain gauges. In addition to radar reflectivity images from individual radars, in the United States all of the weather radar data have been integrated into mosaics that show weather patterns and distribution of rainfall.[9]

Satellite estimates of precipitation are difficult to make and cannot be viewed as surrogates for ground-based precipitation recording stations. Thermal sensors on satellites measure the temperatures of cloud tops, especially for rain-producing convective storms, and record patterns of cold cloud tops that resemble rainfall intensity patterns, as measured by precipitation gauges on the ground. Although there is still considerable uncertainty in such precipitation estimates, satellite imagery does provide spatial information that complements point source networks of precipitation recording stations. Work is being undertaken to merge data from precipitation recording stations, weather-radar systems, and satellite imagery. Much more research is needed to forge operational precipitation estimation tools from such disparate measuring systems.

3.5. WATER USE

In many river basins, significant quantities of water are diverted for agricultural, municipal, and industrial water use, for hydropower generation, and for diversion to other river basins. In some cases, the water is used and returned with little change in its quality, such as in hydropower generation. In other cases — such as for irrigation — some of the water is transpired by agricultural crops and the water is returned to the source of the water body with degraded water quality. Data on these and other water use categories are compiled in some countries so that an accounting of human induced diversions and water use can be made. The diversions are catalogued by use and by geographical location so that aggregated values of use can be made on political as well as natural boundaries. Water use data may be compiled by numerous agencies at the local, state, and national levels, employing a wide variety of data collection methods.

Generally, water use data collection techniques are site specific or estimated from the characteristics of the use. For example, in the former case, consumptive use of water diversions to and from irrigation projects may be accounted for by measuring water flowing in canals within specific projects. Using the latter approach, consumptive use may also be estimated by using remote sensing to inventory crops in the projects and for calculating the water they transpire from known characteristics of the individual plants that are being grown. Since water use is widespread geographically and covers a broad spectrum of human activities, it is impractical to measure all water use everywhere. Rather, attempts are made to sample water use rigorously according to well-established sampling procedures and to calculate total water use from the sample [5][4]

3.6. OTHER WATER RESOURCES MEASUREMENTS

There are other quantitative measurements of water that are useful to water resources management, but difficult to undertake. For example, evaporation of water to the atmosphere and the transpiration of water to the atmosphere by plants are major components in the flow of water out of river basins; collectively these processes are termed evapotranspiration. Evapotranspiration is one of the most difficult measurements of water resources to make. There are many hydrologic services that measure potential evaporation by observing the decrease in water in a pan that is exposed to the atmosphere and then attempting to relate pan evaporation with actual evaporation.

Commonly, hydrologists attempt to quantify all of the water that moves in and out of an area, namely precipitation, river and ground water discharge, and changes in the storage of water in the area; they then compute evapotranspiration as the remaining term in the water budget. Soil moisture is another water resources parameter because it is related to agricultural growing conditions and the propensity for water to run off or percolate into the soil during precipitation events.

3.7. REAL-TIME HYDROLOGIC DATA

Real-time data are an important component in hydrologic data collection and river basin management. These are data that are current measurements of hydrologic conditions in a basin and provide river basin managers with the status of water resources at selected locations. These data are especially important during extreme hydrologic events, namely floods and droughts, when managers must manipulate water-control facilities throughout the basin and issue public safety information to radio and television stations, the police, civil defense organizations and others.

3.7.1. *Real-Time Technologies*

There are several technologies in common use that convey data from data collection locations to central data repositories. They can be used manually or automatically, and fall into two categories: land line and radio telemetry. In the land line case, telephone lines are installed at the hydrologic station and are used manually by a station operator or they are connected to automated recording equipment. By telephone line and modem, a telephone call is established between the station and the data repository and the data conveyed to the repository.

This method now is taking advantage of wireless technology which allows wireless telephone technology to be used for placing the telephone call at many locations. The advantage of the land line method is that two-way communications are possible and the central repository can command certain actions at the hydrologic station. A disadvantage is that during times of flooding, telephone lines and switchboards in communities near rivers may be flooded and communications disrupted.

Radio telemetry takes many forms. For short-distance communication of tens of kilometers in areas where topography is not too severe, line-of-sight radio is in common use in reporting real-time hydrologic data. For example, networks of these stations provide local emergency management officials with current information on water levels and precipitation in areas of the United States that are prone to flash floods.

Where there is a need to collect data over large geographic areas, hydrologic services are increasingly using earth satellites to relay data from remote hydrologic stations. There are two classes of earth-orbiting satellites that are used to relay environmental data: polar-orbiting satellites and geostationary satellites. In the former case, CLS Argos operates on a commercial basis the Argos system, a French communications system that uses United States polar orbiting satellites.[11]

Argos is a global satellite-based location and data collection system dedicated to studying and protecting the environment that is operated cooperatively by the France and the US. Users of the Argos system operate small, battery-operated radios, known as Data Collection Platforms (DCPs), that are connected to environmental sensors. In the case of hydrology, DCPs telemeter hydrologic data whenever the satellites are above the horizon at the DCP site and within radio range. The frequency of data transmission is very dependent on the latitude of the DCP. Opportunities for satellite communications are much more frequent in higher latitudes because the satellites' orbits converge in polar areas.

Geostationary satellites are much in use for the collection of real-time hydrologic data, particularly in the Western Hemisphere. In the United States for example, the National Oceanic and Atmospheric Administration (NOAA) operates two Geostationary Operational Environmental Satellites (GOES) that are in geostationary orbit about 36,000 kilometers above the equator. At this altitude, satellites that are above the equator in a west-to-east orbit move synchronously with the rotation of the Earth and appear to be fixed above the equator. Each GOES satellite views nearly 50 percent of the surface of the Earth and collectively they view all of North and South America and large portions of the Pacific and Atlantic Oceans.[12]

At locations on the Earth's surface near the periphery of the satellite's viewing area, radio communications may be impossible because the satellite is close to the horizon and communications may be obstructed by local topography. Otherwise, communications between DCPs and a satellite are possible 24 hours per day. Most

DCPs collect data from sensors as frequently as every 15 minutes, then transmitted the collected data every 3 or 4 hours to a GOES satellite. Many are equipped with an alert feature, which allows the DCP to transmit more frequently when the water level at the site exceeds a predetermined flood level. Through the GOES system, data from around 10,000 DCPs are provided to data repositories in real time by a variety of means.

The most common method is one wherein all DCP data that are relayed through the two GOES satellites are immediately transmitted from the National Environmental Satellite Data and Information Service (NESDIS) at the Command and Data Acquisition Center in Wallops Island, Virginia, to a domestic communications satellite for relay to data users nationwide. Data that are transmitted through a domestic communications satellite permits users to access data via relatively inexpensive receiving stations.[13]

Geostationary meteorological satellites similar to the GOES satellites are operated by the European organization Eumetsat, which operates Meteosat, and the Japanese Meteorological Agency which operates the Geostationary Meteorological Satellites (GMS).[14]

A radio telemetry system in widespread use in the United States for hydrologic data collection is the SNOTEL system. SNOTEL stands for Snow Telemetry and is operated by the Natural Resource Conservation Service (NRCS). This system takes advantage of transitory ionized trails in the atmosphere that result from the passage of micrometeors that rain upon the earth continuously. These trails reflect radio waves and permit brief radio communication between widely separated points. This technology, known as meteorburst technology, is used by NRCS to monitor snow conditions at about 500 sites in the mountainous western part of the United States.

Because the ionized trails in the atmosphere are transitory, a central data repository that is polling a remote hydrologic station may not be able to establish communications and extract the hydrologic data immediately upon demand. Nevertheless, the NRCS has more than 20 years of experience in operating the SNOTEL system with meteorburst technology and has good performance statistics on this type of communications and data-collection technology. SNOWTEL serves the important purpose of monitoring the accumulated snowpack in the mountains of the American West, which is the source of much of the water for agriculture and municipal water use.[15]

4. Geographic Information Systems

In recent years, computer-based mapping systems have become powerful technologies to analyze and display natural resources data. All of the types of maps cited in the Water Resources Information section of this study can be digitized and stored in a computer database so that aggregated layers of information are available in the computers. Broadly speaking, Geographic Information Systems (GIS) can be used to analyze and display map-based data. To tap the aggregated information, one could then query the computer and, for example, determine the number of people who live within a specified distance of sources of ground water contamination. Or, if weather data also were summarized and stored in the computer, one could operate atmospheric diversion models and estimate the number of people who could be affected by condensation from fossil fuel electric generating evaporative cooling facilities.

As regards the displaying of data, one could conduct flood frequency studies from stream-discharge data from all of the river discharge stations in a basin, compute the magnitude of a flood that one could expect every 25 years, then display the information in spatial format. If large scale topographic data and the locations of hospitals, schools, and other public facilities were also in the computer, one could map the extent of the 25-year flood in a community and identify all the public facilities that would be at risk during the flood. Moreover, all of the analyses discussed above could be displayed spatially on maps that would be made available to river basin managers, civil defense officials, and other government and business officials. USGS water use data commonly are portrayed on state or national maps that define counties and use color to code interval ranges of water withdrawals. Historical data from numerous water quality monitoring stations in California have been analyzed for trends and the results shown on the maps as circles (no trend) or up or down arrows (increasing or decreasing trends).[16]

5. Data-Collection Standards and Support

In order to collect historical data, hydrologic services operate data-collection networks over many decades, during which time technologies, staff, and budgets change. Throughout the data collection period, in order to determine whether the hydrology has changed during the time that data have been collected, hydrologic services must maintain the quality of their data collection programs and provide support to data collection staff.

In order to collect data that are of consistent quality throughout the long period of data collection, hydrologic services operate their data collection networks consistently in accordance with established standards. These standards, which are published in official documents, are constantly referenced by staff as they collect the data and by others as they use the data. For example, a hydrologic service will publish a document that states how a river discharge measurement is to be made. The report will cite the characteristics of the current meter that is used, how many current meter measurements must be made in each vertical section of a river's cross section, and how many horizontal sections must be measured. The document also will document how the river discharge measurement is computed and stored in central data repositories.

If current meter data are collected and used consistently to compute river discharges at specified sites, then staff turnover, improvements in current meter technology, and computer-assisted discharge computations can be introduced to the data-collection program with no loss of comparability of river-discharge data.

Successful hydrologic services operate quality assurance programs to validate that data are collected and reported according to published standards. For example, a team of hydrologists from headquarters or from regional and field offices of a hydrologic service will conduct an inspection visit to another office, where they will observe data-collection and analysis practices, interview staff, inspect current meters and other equipment, review current meter data and river discharge data, and document their findings.

These quality assurance inspections and the reports that they produce are intended to verify that data collection programs are operated properly and to identify any problems that need solution. One of the most effective ways to assure quality data is to make the data available for widespread use by scientists and engineers, both within and outside the organization that collects the data. In their use of the data and by comparisons of the data with other data stations in the basin, spurious values, questionable trends, and other failures in the data collection system tend to be discovered.

Other quality assurance programs affirm that data, maps, and analytical reports meet established technical, policy, and editorial standards of the organization before they are published and released to the public. In successful hydrologic services, quality assurance programs to review data collection, analyses, and publications are assigned to relatively independent offices in the organizations, and these offices pursue the responsibilities vigorously.

Staff are the most valuable resource of a hydrologic service and it is critical to provide training and technical support to them as they conduct data collection programs. Training of staff may take many forms, from on-the-job training for new staff to formal university degree programs for more experienced staff. Additionally, many hydrologic services maintain their own training facilities where staff take short courses on data-collection, report writing, hydrologic modeling, instrumentation, and other topics throughout their careers.

Among the most critical support programs for hydrologic data collection are those for water quality laboratories and instrumentation. As discussed above under water quality data collection, water quality laboratories provide analytical services to the data collection program. In addition to analyzing water samples to determine the quality of water and forwarding the data to central data repositories, a laboratory may provide sampling bottles, chemicals to preserve samples, and other technical assistance to data collection staff. In cooperation with the training program above, laboratory staff themselves may participate in training programs, research improved analytical techniques, and they may participate with other water quality laboratories in quality assurance programs.

Hydrologic data collection relies extensively on instrumentation. Such instrumentation includes current meters, water level sensors, data loggers, DCPs, portable and automated water quality monitors, and water quality samplers. Some instrumentation support programs may maintain inventories of commonly required instrumentation so that instrumentation needs of field offices can be met promptly.

The instrumentation support program may design data collection instruments, procure design and manufacturing services from the private sector, or buy off the shelf instruments from the private sector. In the latter case, off-the-shelf instruments may be tested by the instrumentation support program to assure that the instrumentation meets data collection standards. Furthermore, the program may be responsible for periodic testing of all current meters, to assure that the meters measure water velocity properly over the designed range of measurement. Whether designed by the instrumentation support program or by the private sector, the program must assure that the instrumentation meets the organization's published data collection standards.

Many hydrologic services conduct research that leads to improved data collection technologies. Such research may be oriented towards developing new hydrologic measurements — such as inventorying the biota in a river as a standard of water quality — or developing new laboratory analytical techniques. The research may provide new instruments for measuring, recording, and transmitting hydrologic data, or new applications of emerging technologies, such as GIS, that allow hydrologic data to be transformed to water resources information.

6. Sources of Information on Hydrologic Data-Collection

The World Meteorological Organization compiles a broad variety of hydrologic information via the HOMS program which provides access to technical documents drawn from international sources.[17]

Another principal source of information on methods of hydrologic data collections is a handbook of recommended methods that has been prepared cooperatively by agencies of the United States Government. (U.S. Geological Survey, 1978) Some of the eleven published chapters in the handbook were published as early as 1978 and updated from time to time. The document is entitled "National Handbook of Recommended Methods for Water-Data Acquisition."

The topics of the eleven chapters of the handbook and data of preparation are shown in Table 4.

TABLE 4. Chapter, titles, and year of publication of *National Handbook of Recommended Methods for Water-Data Acquisition*

Chapter	Chapter Title	Year of Publication
1	Surface Water	1980
2	Ground Water	1980
3	Sediment	1978
4	Biological and Microbiological Quality of Water	1983
5	Chemical and Physical Quality of Water and Sediment	1982
6	Soil Water	1982
7	Basin Characteristics	1978
8	Evaporation and Transpiration	1982
9	Snow and Ice	1979
10	Hydrometeorological	1980
11	Water Use	1996

Additionally, the USGS publishes a series of manuals describing procedures for planning and conducting specialized work in water-resources investigations. The Techniques of Water Resources Information (TWRI) and other documents can be searched for at http://www.usgs.gov/usgssearch.html. Many are listed at http://txwww.cr.usgs.gov/reports/wdr/95/SW/techpublications.html and are available for sale by the USGS, Branch of Information Services, Box 25286, Federal Center, Denver, Colorado 80225.

The U.S. Environmental Protection Agency also maintains a Water Resource Center as a central point of contact for the general public. The Center contains scientific and technical documents from EPA water programs and coordinates with other national clearinghouses and can be contacted via http://www.epa.gov/OST/rescnter.html

Internet sites listed at http://bordeaux.uwaterloo.ca/enviro.html provide access to a variety of Canadian environmental organizations, whereas http://www.cna.gob.mx/ provides access to the Mexican national water agency, Comision Nacional del Agua.

7. Summary

Data on a modest number of relatively easy measurements are the basis of highly useful water-resources information that are needed by river-basin managers. These data include water level, river discharge, precipitation, water use, and a few basic water quality measures. Historical records of these data provide scientists and engineers with a basis for estimating the recurrence of extreme hydrologic events, which river-basin managers use for planning. Real time hydrologic data provide river-basin managers with information on the state of the basin, which they use for real-time decision making.

In order to collect these data, hydrologic services must:

- adopt a data-collection standard,
- operate quality assurance programs to affirm that the data are collected properly, and
- provide support programs to:
    - train staff,
    - supply water-quality analytical services,
    - maintain instrumentation, and
    - conduct research to improve data-collection technologies.

If numerous data-collection agencies in an international river basin collect similar data, but do not adhere to data-collection standards, quality assure their programs, or provide support to their staff, then data collected by the several agencies will not be comparable and many information needs of river-basin management will not be met. There are many hydrologic services that successfully operate historical and real-time data-collection programs and their publications, experience, and staff provide a reservoir of information for organizations that contemplate establishing new data-collection programs. Many of these publications and staff experience are becoming available on the global computer network known as the Internet.

Notes

1. Buchanan, T.J. and Somer, W.P. (1968) Stage measurements at gauging stations, Techniques of Water Resources Investigations, Book 3 Chapter A7, 28 p.

2. —— (1969) Discharge measurements at gauging stations, Techniques of Water Resources Investigations, Book 3 Chapter A8, 65 p.

3. Kennedy, E.J. (1984) Discharge ratings at gauging stations, Techniques of Water Resources Investigations, Book 3 Chapter A10, 59 p.

4. Laenen, Antonius (1985) Acoustic velocity meter systems, Techniques of Water Resources Investigations, Book 3, Chapter A17, 38 p.

5. U.S. Geological Survey (1990) National water summary 1987-Hydrologic events and water supply and use; U.S. Geological Survey Water-Supply paper 2350, 553 p.

6. —— (1991) National water summary 1988-89-Hydrologic events and floods and droughts; U.S. Geological Survey Water-Supply Paper 2375, 591 p.

7. —— (1995) The strategy for improving water-quality monitoring in the United States, Final report of the intergovernmental task force on monitoring water quality, U.S. Geological Survey Open-File Report 95-472, 25 p.

8. Wahl, K.L, Thomas, W.O., Jr., and Hirsch, R.M. (1995) The streamgauging program of the U.S. Geological Survey; U.S. Geological Survey Circular 1123, 22p.

Some of the references embedded in the text refer to the World Wide Web Universal Resources Locators (URL) on the Internet. The latter references are especially valuable because public and private organizations that make their information available on their so-called "home pages" on the Internet, continue to refresh them with current information. The latter references generally are of the form "http://water.usgs.gov" and "http://www.nws.noaa.gov" where the first seven characters in the URL are "http://" These latter are, respectively, the URL's of the U.S. Geological Survey and National Weather Service home pages.

There also are so-called "search engines" on the Internet that allow a user to search for sources of information. A search of the term "hydrologic" for example returned more than 15,000 sources of hydrologic information, ranging from universities and government agencies to private sector companies. However, the Internet is a very dynamic source of information and the URL's shown will be changed from time to time and other URL's added as additional organizations offer their information on line.

Please Note:

- The author does not endorse any of the commercial sources of information that are shown here, nor are the sources completely representative of the commercial products available.
- Because meteorology and hydrology comprise most of the hydrologic cycle, these data commonly are referred to as hydrometeorology data.
- Water stage is another common term for water level.

- Water level is an important measure in estuarine environments where astronomical and wind-driven tides join river discharge as driving forces for water-level changes.
- Concentration loads are computed by multiplying the river discharge by the constituent concentration.
- Inquiries about recommended methods should be directed to the Chief Hydrologist, U.S. Geological Survey, MS 409 National Center, Reston, Virginia 22092, USA
- This chapter is on line at http://h2o.usgs.gov/public/pubs/chapter11/index.html

---

[1] This information may be found on the Web at http://salmo.cqs.washington.edu/~wagap/nm/wildlife/5hydro.htm

[2] This information may be found on the Web at http://cdec.water.ca.gov/

[3] Real-time hydrologic data is treated as a separate issue at the end of this section. Network design of hydrologic stations is an important consideration (U.S. Geological Survey, 1995a), but, because of the limited scope of this paper, the subject not considered herein.

[4] Examples of information on commercial vendors who supply water-level and other environmental measuring and data-logging systems can be found at: http://www.green-world.com/4210.htm http://www5.electriciti.com/h2ogage/watrgage.html http://www.sutron.com/products/sensors/default.htm, http://www.csluk.demon.co.uk/ http://www.handar.com/html/sensors.htm, and http://ourworld.compuserve.com/homepages/ESPL/tgif.htm

[5] Information about the use of ADCP's to monitor river dynamics and sediment transport is described in http://w3g.gkss.de/G/GSS/highlight/highlight1.htm; more information on ADCP's from a commercial supplier can be found at http://www.adcp.com/

[6] Commercial suppliers of such equipment include http://www.dbed.state.md.us/Environment/83.html

[7] The home page of the USGS National Water Quality Laboratory is at http://wwwnwql.cr.usgs.gov/

[8] See information on the United States Environmental Protection Agency's acid rain program at http://www.epa.gov/docs/acidrain/ardhome.html

[9] See, for example, http://www.intellicast.com/weather/usa/radar/; locations of weather radar systems in Mexico can be found at http://www.cna.gob.mx/radares/radares.html

[11] See for example, http://www.newspace.com/Industry/list/company/1066.html

---

[12] Current imagery from GOES-8 and GOES-9 can be found at http://www.goes.noaa.gov/index1.html

[13] For a more exhaustive discussion of the technical characteristics of the GOES system, refer to: http://psbsgi1.nesdis.noaa.gov:8080/EBB/pubs/TR40/21.html. For information about the use of GOES to collect data from more than 4,200 hydrologic stations by the US G S can be found at: http://h2o.usgs.gov/public/pubs/circ1123/overview.html). Public access to these real-time data can be obtained through http://water.usgs.gov/public/realtime.html. An example of river gaging station water-level and discharge data in the State of Pennsylvania can be found at http://wwwpah2o.er.usgs.gov/rt-cgi/gen_stn_pg?station=01570500. Access to National Weather Service River Forecasting Centers that use these real-time data for river and flood forecasting can be found at http://info.abrfc.noaa.gov/rfc-wfo.html

[14] See http://www.belspo.be/telsat/meteo/meax_dcs.htm , whereas information about the GMS can be found at http://se.eorc.nasda.go.jp/GOIN/JMA/htdocs/JMA-NET1.html

[15] See http://wrcc.sage.dri.edu/snotel_climate.html

[16] Numerous on-line maps of sources of nutrients in water in the United States can be found at http://wwwrvares.er.usgs.gov/nawqa/nutrients/maps.html

[17] See http://www.wmo.ch/web/homs/subsecs.html#C

DATA STANDARDIZATION

WILLIAM J. SHAMPINE
U.S. Geological Survey
P.O. Box 25046, MS401
Denver Federal Center
Lakewood, CO 08225 USA

Abstract

Sharing data effectively requires data that are consistent and comparable over time, of known quality, and both available and accessible. Common problems associated with historic databases are the lack of proper documentation, lack of ancillary data, and a changing ability to store documented records. The single most important technique to evaluate masses of data is to put them into some kind of a graphical or pictorial form. Collecting data of known quality requires the use of established standard data collection and analysis methods, careful documentation of the work done, and preparation and implementation of a Quality Assurance Plan. Quality assurance (QA) and quality control (QC) activities effectively address many of the potential problems organizations might have with knowing the quality, consistency and comparability, and availability and accessibility of data stored in various databases.

Key Words:

data evaluation, data quality, quality assurance, quality assurance plan, quality control, standards, water quality

1. Introduction

Life as we know it requires water, which is a fixed commodity on the planet. As local and global populations increase, with the commensurate growth in water use and the attendant degradation of the water, it becomes increasingly imperative that humankind manage the available water resources effectively. This successful management of this vital task requires adequate and accurate data about those resources. Although water managers want data in order to make wise decisions, *it must be understood that they will make their decisions when they have to, with or without data, and whether or not the data available are sufficient or accurate.*

It is the responsibility of hydrologic service organizations to provide the appropriate type and quality of data needed by water managers. Since hydrologic data are

T. Naff (ed.),
Data Sharing for International Water Resource Management: Eastern Europe, Russia and the CIS, 45–59.

expensive to collect, it would be advantageous from a scientific and cost-effective viewpoint if all of the data available for a given hydrologic unit could be shared, regardless of the political boundaries from which those data were collected. Sharing data clearly would maximize the information available to a manager at a minimum cost to any single organization. Sharing data is a sound concept technically, although periodically it may not be politically feasible to do so. Clearly, the concept of sharing data is much easier to advocate than it is to accomplish. Further, assuming political issues can be resolved, there also are many technical and logistical issues associated with sharing hydrologic data that must be taken into account. In order to share data from multiple sources, the following requirements must be satisfied:

- Data of known quality
- Data types and quality that are consistent and comparable over time
- Data that are both available and accessible.

In order to share data effectively, there are two additional related needs:

- A computerized database to store the data
- Software to process and analyze the data.

The addition of a computer system to store and handle the data makes it significantly easier and efficient to deal with the typically large volumes of hydrologic data collected today. The data can be processed by hand rather than by computer, but that method is a very time-consuming process. This paper will focus only on the quality, consistency, and availability of the data and will not include information about databases.

2. The Need for Data of Known Quality

For a distributed database system to be useful and effective, it is critical to know the quality of the data. It is insufficient to say the data must be of "good" quality because "good" is a relative term and means different things to different people. What is "good" quality data for one person may be "poor" quality data for someone else. It is better to "know" and make known the actual quality of the data. With such knowledge, each concerned person can then evaluate the usefulness of the data for individual purposes.

The quality of data are defined as being *known* when the precision and bias of those data are known. The question then arises, how are the precision and bias of the data known? Standardized, published procedures and methods of data collection and analysis always include an evaluation of the precision and bias produced by using those methods. Thus, the precision and bias of data are "known" if they have been collected following accepted and documented procedures and if the conditions and circumstances associated with the data collection and handling are described and documented. All too often, however, data are collected by non-standard methods or there is no information available describing how the data were collected. In these circumstances, a variety of problems can arise in efforts to interpret these data.

Knowledge of the data quality is particularly important in the area of water quality because small differences in sample collection techniques or analytical methodology can make significant differences in the results. When hydrologic data are collected over a long period of time, it is possible to experience complications even when following accepted and documented procedures because the "accepted" methods will change over time along with advancing technological changes.

For example, a Canadian researcher was collecting phosphorus data on some lakes in Canada in the late 1960's. In this particular instance, the researcher had some historic data collected in the 1940's and 1950's to supplement the data just collected in the 1960's. The phosphorus concentrations in the samples collected were about 1.0 mg/L in the 1940's, about 0.4 mg/L in the 1950's, and about 0.05 mg/L in the 1960's. The conclusion derived from a plot of these data was that the phosphorus concentrations in the lake were decreasing with time. In fact, this conclusion was wrong. Fortunately, the researcher had access to information on how all of the samples were collected and treated prior to analysis (the scientists in the 1940's and 1950's had documented their procedures). All of the samples had been collected by the same method: by dipping a bottle into the lake.

However, the samples were unfiltered in the 1940's, filtered with standard filter paper in the 1950's, and filtered through 0.45 micron paper in the 1960's. This information inspired an idea. The researcher did some additional testing and was able to determine that the phosphorus concentration probably was not related to time, but to the nominal pore space of the filter system. Apparently, rather than being dissolved in the water, the phosphorus was attached to the cell structure of the biota, and thus the change in concentration over time was caused by a change in the amount of biota left in the sample after filtration. The analytical procedure was such that any biota remaining in the sample aliquot would be destroyed and it was the phosphorus contained in it that was being measured. This example demonstrates the value of having documentation on the data collection procedures employed in collecting samples when trying to analyze the data.

The measurement of pH is another water quality example of the need to know something about the conditions of measurement. The pH of surface water varies in a cyclical pattern in response to the photosynthetic activity of the biota in the same water. Therefore, it is necessary to know the time of day, water temperature, and cloud cover conditions at the time the measurement is made. If the data were collected consistently at about the same time, the pH, barring any other outside influence, would probably be relatively consistent in value. However, if the measurement were made at varying times throughout the day, the pH content could vary as much as two pH units as a direct result of photosynthetic activity and have nothing to do with pollution, which might otherwise be indicated or at least suspected.

Problems with hydrologic data of unknown quality are not always complex, although they may be difficult to identify. As a third example, a hydrologist, while evaluating surface water flow data collected over several years, noticed that the high flow data from one station did not correlate well with the same flow data from other stations in the basin. A detailed investigation eventually revealed the simple reason for this apparent anomaly. To facilitate making high-water measurements during flood conditions, a technician had painted distance markers on the bridge to eliminate having to mark the sections every time a measurement was made. But, these markers had been erroneously painted at four-foot intervals instead of five feet apart which was the normal spacing used by the analysis office. Assuming that the markers were placed at five-foot intervals without having actually checked their spacing, and thus being unaware of the error, the hydrologists were calculating the measurement computing areas based on five-foot rather than the actual four-foot width separating them. Since these bridge markings were used only for high flows, the upper part of the rating curve consequently gave discharges about 20 percent too large.

The point of these examples is that not knowing the quality of the data being used — that is, the precision, bias, and method of collection — sometimes leads to erroneous interpretations and conclusions about those data. The user of the data must always be very careful.

3. The Quality of Historic Data

One of the larger challenges for a hydrologist is to evaluate the quality of data available in large or multiple data sets that have been collected over many years by many different people. Typically, little is known about how such historical data were collected. Moreover, all too commonly, the data collection techniques were inconsistent over the period of record, which probably made the data incompatible. Between the two key requirements for establishing the quality of any data — standard methods and documentation — the most common failing is a lack of documentation rather than the application of standardized methods.

For short-term data programs, there are, typically, neither sufficient incentives nor some form of compulsion to ensure that an individual or organization will document work activities beyond the level of personal notes. The pattern often followed in long-term monitoring programs is inconsistently maintained documentation; the level of record-keeping of collected and stored data may be sufficient at the beginning of the program, but often is not kept current as operation becomes routine. Thus, as new technology results in changes in some of the field and laboratory methodologies being used, the paper trail documenting these changes may be lost.

Two other problems that plague the maintenance of historic hydrologic data sets are a lack of ancillary data and alterations in the capacity to store documented records. Ancillary data, which describe sampling conditions and circumstances, might include such information such as cloud cover, air temperature, water color and clarity, odors, well screen locations, casing type, stream turbulence, presence of algae, sample filtration techniques, etc. In years past, most hydrologists either did not realize the real value of this type of information or they had no convenient way to deal with it. So ancillary data went largely uncollected. Even when historic ancillary data have been collected, they have tended to be poorly associated with the relevant environmental data to which they apply. In many instances, much of the ancillary data was recorded as personal notes on scraps of paper, which may have ended up in a carefully stored box somewhere in a basement, or at best, in a field notebook, which also may be in that box in the basement. In either case the ancillary data could not be conveniently applied to the data in a database.

3.1. CHANGES IN THE HANDLING AND STORAGE OF DATA

Changes in the treatment of data have not been confined to the collection and analysis of hydrologic data. Significant alterations have also been introduced into the handling and storage of data. A couple of decades ago, most of the collected documentation and ancillary data were prepared and stored in a paper format. Paper files are generally difficult and time consuming to work with because of the way they must be accessed and because of having to sort through volumes of information to locate the specific items needed. Furthermore, because paper files are incompatible with the computerized data handling and storage techniques in common use today, a tendency has developed to ignore the old, difficult-to-work-with paper data, and focus on the new, easy-to-work-with electronic data.

Over time, new technologies have overtaken even the improved computerized handling and storage of data. In the early years of computer technology, storage capacity was very limited. This limitation resulted in giving priority to the storage of raw hydrologic data with only small amounts of ancillary data stored in association with the raw data. However, continuous technological advances in computers have resulted in giving them virtually unlimited storage capacity. This change makes it easier to store in electronic format all of the combined information desired. Ironically, the computer capability today has caused the nature of the problem of ancillary information to shift from one of not enough to one of too much information. Presently, in some situations, it appears that some information is stored more because the capacity is there rather than because of real need.

3.2. WATER QUALITY MONITORING PROGRAMS

The number and variety of technical problems associated with trying to use historic data vary in response to the type of hydrologic data being used; among such data, water quality data is the most complex. Owing to frequent technological changes it is particularly difficult to predict trends using data from long-term water quality monitoring programs. The underlying principle of all water quality monitoring programs is to collect data over a period of time in order to be able to evaluate changes in the data and, if possible, to develop information on trends. To accomplish this objective, all of the data being evaluated must be of the same type, the quality should be consistent, and the data comparable throughout the period of record.

Adherence to these criteria may appear to be simple but in reality that is not the case. The reality is that where water quality data is concerned, the ability to develop information on trends over a long time has proven to be almost impossible. In a 1991 paper on detecting and estimating trends in water quality, Hirsch, Alexander, and Smith address the need for consistent data and describe some of the many factors impacting data consistency and comparability. In discussing sample collection and analytical methods they write:

> It is assumed for purposes of this examination that one or more sets of data, which were collected over a period of years in a consistent and reliable manner, are available to the investigator. This means that the rules for the timing of sample collection must be known (convenience

> sampling is not acceptable), the methods of sample collection, handling, shipment, preservation, laboratory measurement, and data reporting conventions (rounding and reporting limits) must be constant over the period of record. There can be exceptions to this requirement of constancy. Specifically, if changes have been documented to have no effect on the resulting data, or if changes result in known biases and these biases are subsequently corrected in the data undergoing analysis, then the procedures described here may legitimately be used to examine the data for trend. [3]

Because it is easy to understand the need, the collection of the same type of data is, normally, not a problem for water quality monitoring programs. The need for data of a consistent and comparable quality also is relatively easy to understand, but actually is difficult to accomplish. Because of the inherent methodological changes in processes associated with advancing science and technology, it is very difficult to ensure consistent and comparable data quality over a long period of time, even when the desire to do so exists.

Constant technological advances in the collection and analysis of environmental data commensurately enables the collection of more accurate data. This type of change makes it very difficult to interpret many types of historic data because the environmental samples collected after the introduction of each new technology may not represent the same statistical population as the samples collected prior to the change. One has to be very careful in comparing historic water quality data because the sample collection and analytical methods have changed significantly over the years; this circumstance has changed the precision and bias of the data.

Despite the potential problems, most long-term data collection programs will, nevertheless, adopt new techniques as they are developed because the pressure to collect more accurate data is very high. The new potential for working with data of varying levels of precision and bias, and of varying statistical populations, make it critically important for every program to document in detail how the data are being collected and to detail all of the changes that occur throughout the life of the program. This level of documentation, if readily available, would not only allow the scientists managing the program to evaluate properly their own data, it would also allow all future users of the data to evaluate the data for their own use.

3.3. TECHNIQUES FOR EVALUATING HISTORIC DATA

In endeavoring to evaluate the quality of historic data, analysts are often faced with a mass of both data, in both paper and electronic form. The challenge for analysts who would like to use the historic data is to decide if the data meet their current needs. Are the data from the same statistical population? Are they comparable? It may be impossible to determine “truth”, but it is necessary to make some evaluation of the data quality before trying to interpret them.

The single most important procedure for evaluating masses of data is to put them into some kind of a graphical or pictorial form — geographic displays, *x-y* plots, charts, diagrams, etc. Surface water and groundwater data in particular should be plotted geographically by hydrologic units or by aquifer systems and the resulting displays searched for obvious inconsistencies. If the collecting organization is known, the data also should be plotted in a manner that shows the differentiation. This type of differentiation will sometimes demonstrate that data from Organization A may plot "higher" or "lower" than the data from Organization B.

It is suggested that the first step in evaluating historic data is to convert the data into an electronic form if possible. Although not an absolute requirement, such a conversion would certainly make the data more convenient to work with and thus make analysis of the data significantly easier. Even though data entry into electronic form is a time consuming process, there are three great advantages which may make the investment of time worthwhile: (1) data handling chores are greatly simplified; (2) the user is able to analyze the data employing a variety of graphical and statistical packages with relative ease; (3) future users will be able to work with the data with much greater facility. If left in paper format, the same cumbersome data handling process must be repeated each time there is a need to look at the data.

After transferring the data and information into electronic form, the next step is to prepare graphic representations of the data and displaying them in a variety of ways. Each graphic will yield slightly different information. The objective of the displays is to search for obvious data "outliers", that is, data that seem not to fit with the other data. It is important to remember that outliers may or may not be erroneous data. These "special" data require a close look and evaluation, which may show them to be (1) incorrect or (2) caused by a hydrologic condition not presently understood. Equally, even though all of the remaining displayed data appear to be compatible does not prove they are comparable, but it certainly is a good indicator of relative consistency.

A good, basic reference book for graphic displays is the work of Velleman and Hoaglin who describe a series of nine exploratory data analysis (EDA) techniques which have added a new dimension to the way analysts approach data [8]. EDA applies both hand-plot and computer-plot techniques to uncover features concealed among masses of numbers. Their techniques include *Stem-and-Leaf Displays, Letter-Value Displays, Boxplots, x-y Plotting, Resistant Line, Smoothing Data, Coded Tables, Median Polish, and Rootograms.* In addition to demonstrating how these techniques can be applied by hand, they also include the computer source code in both BASIC and FORTRAN programs, allowing users to enter the code on their own computer if desired. The computer information in this reference may be considered old, but the science it contains is excellent and still valid.

Water quality data are a bit more complex than surface or groundwater data because of the large number of parameters involved and because of the sensitivity of the quality measurements to a wide variety of factors. While geographical displays are useful, other types of plots can be very helpful. For example, useful data plots from a given site could include ion-ion, ion-specific conductance or total dissolved solids, and time series. Calculating and plotting the analytical charge balances, Stiff Diagrams, and Trilinear Diagrams can also help identify potential data problems. Descriptions of all of these techniques for water quality data analysis are described in Hem [2]. All of these plots can be done by hand, but it is much easier and faster if they are carried out by computer. (It should be noted that Current hydrology-oriented journals advertise several relatively inexpensive commercial software packages available for PC computers that are designed to generate these types of plots quickly and with a minimum of difficulty.

In addition, some spread sheet software, such as Microsoft EXCEL, is an excellent tool for drawing *x-y* plots.)

After identifying the data outliers, one must then decide what to do with them. Simply discarding all data outliers carries the risk of discarding valuable information. It must never be assumed that because one does understand the data it must be wrong. Data that do not fit the norm should be treated cautiously, but only with an awareness that not all data will conform to normal expectations. For example, a plot by the author of miscellaneous stream temperatures measured in Indiana showed what looked like an outlier temperature of 9 degrees Celsius measured in February for a stream in an area with severe winters. The expected value should have been about 3 degrees Celsius, the historic average at that time for streams in the area. However, after looking at the daily air temperature records in the same vicinity and time frame, it became apparent that there was a significant period of unusually warm weather that particular February. The conclusion was that the "outlier" value was likely correct and should not be discarded. Sometimes, one must become a hydrologic detective in order to evaluate the data accurately.

4. How to Produce Data of Known Quality

The means by which hydrologic data were collected in the past cannot be changed, but it is possible to control how hydrologic data can be collected in the future. It is always advantageous to collect data in such a way as to reveal the quality of the data. One of the chief benefits of such an approach to data collection is that anyone who has a need for and accesses those data would know their quality. Another important payoff is that knowledge of data quality significantly improves the ability to share data effectively.

It is worth reemphasizing that the collection of data of known quality requires (1) the use of established standard data collection and analysis methods, (2) careful documentation of all of the work done, and (3) preparation and implementation of a Quality Assurance Plan.

4.1. STANDARD METHODS

The advantage of using established standard data collection and analysis methods is that they produce data of known quality with established levels of precision and bias. They are published and readily available for any potential user of the data. Sometimes, however, hydrologic conditions prevent the use of published standard methods. If it is not possible to use standard methods, it is critical that whatever methods are used are carefully documented and kept in a form and location that will always be accessible.

4.2. DOCUMENTATION

Documentation is probably the single most important requirement for knowing the quality of data being collected. Literally everything about an investigation or data collection program needs to be carefully documented. Information on the methods and analytical tools used, and especially the assumptions that frame the program must be exposed; this information would be invaluable to analysts in the future. There is a strong tendency to forget or exclude from the documentation package the underlying assumptions made in a program. Such oversights can create enormous problems for those trying to use the data.

Water use data provide excellent illustrations of potential problems. For example, the technique used to calculate water used by chickens within some geographic boundary typically involves multiplying the number of chickens in the area by the amount of water used by a single chicken for some specified time period. Since the amount of water used by a single chicken for a given time period is typically an estimate, it is critical that the estimate used today is the same one to be used in the future, unless there is a documented change in the water use of a chicken. To do otherwise would yield artificial changes in the data. The only way to maintain the consistency of data in the database is to document the assumptions made so that the same assumptions can be used to develop new data in the future.

Another requirement for the preparation of useful documentation is assurance of the availability and accessibility of the documents themselves. This factor may seem almost too self-evident and simple to cite as an important necessity, but it is commonly a major flaw in otherwise successful programs. Not infrequently, both paper and electronic data archives are often difficult to locate or are not easily accessible after long periods of time have elapsed since the archives were created.

Documentation and record-keeping relative to the data archives are chores that many persons tend to postpone because these chores are considered to be drudge work and not as challenging or interesting as other aspects of the project itself. The project managers often delude themselves with good intentions of eventually and catching up with backlogs of "paper work" just as soon as a few days of spare time become available; in time intentions become confused with action. Enormous amounts of information have been lost as a result of problems with storage information systems. Much valuable information has been discarded over the years because no one in the organization who might have used it knew anything about its existence.

The best solution to the problem of data availability and accessibility is for the program manager to plan for periodic record keeping and documentation updating and to schedule time for these chores. If the documentation is not kept current, significant amounts of information may be lost as the personnel change during the lifetime of the program.

5. Quality Assurance Plans

Preparing and documenting a Quality Assurance (QA) Plan will help ensure that planned activities are conducted as designed. The two major objectives of a QA Plan (QAP) are to (1) state clearly the expectations of management for the staff and (2) help ensure that planned activities occur as scheduled. By describing a process to help meet these two objectives, a QA Plan functions as a system of "checks and balances." Adherence to the process laid out in the plan prevents unpleasant surprises and helps to ensure consideration of all possible actions. It may be, for example, a manager chooses not to take some specific action, but by following the processes described in the plan, the action he chose not to take would have been a conscious, deliberate management decision, rather than simply something that was overlooked.

Quality Assurance and quality control (QC) address problems encountered by organizations relative to the matters of knowing the quality, consistency and comparability, and availability and accessibility of data stored in various databases. An effective QA program could extend the utility of the data from an organization that collected the data to the entire scientific community. Environmental data that has undergone proper quality assurance would allow anyone to evaluate the data relative to his or her own needs.

Although QA and QC are terms that often are used interchangeably, there are distinctions between them. QA is a management tool that addresses all aspects of a program and establishes an operational framework to help ensure all of the factors affecting that program are considered. QA can be defined as: "All those planned or systematic actions necessary to provide adequate confidence that a product or service will satisfy given requirements of quality." QC is a "worker" tool and represents the day-to-day actions involved in collecting the data — actions like instrument calibration, blank samples, test samples, check lists, and ensuring the use of proper sampling techniques and the application of standards. QC can be defined as: "The operational techniques and the activities used to fulfill requirements of quality."

A QA program should not be considered as a means to improve the quality of environmental data being collected. It is, rather, a means to *document* the quality of the data being collected. The primary objective of the QA effort is to help ensure that planned activities occur as they are planned — no more, no less. If the planning or data collection techniques are flawed, a QA program will not overcome the flaws. Collecting a lot of QC data in association with a poor environmental sample or measurement only yields a poor sample or measurement that is well documented.

The best way to ensure success with a data collection or hydrologic investigation program is to have a written plan that describes in detail the work to be done, and when and how that work will be completed. Along with a work plan, a quality assurance plan is needed which describes in detail the planned actions designed to ensure the work will meet the stated objectives and will be completed as intended. The basic goal of a QA Plan is to provide a means to verify and document the quality of the data being produced. It can be written either as a separate document or as an expansion of the work plan wherein the QA and QC actions are described in close association with the pertinent work elements.

QC data should be collected and utilized as an integral part of the QA effort associated with a sampling program. These data allow for the quality and suitability of the environmental data to be ascertained and evaluated. The specific type of QC data to be collected, selection of the sample sites, and the frequency of collection are critical issues that should be considered very early in the planning stages of a sampling program. Care must be taken to ensure that the QC data are representative of the environmental data to be collected in that they are collected in a time and spatial pattern similar to the environmental sampling.

The QC data should be used regularly as a tool throughout the life of the sampling program. By routinely assessing and evaluating the data as they are collected, rather than leaving them until the end of the program, the program manager can effectively monitor the progress of the work and possibly recognize a need for changes in sampling locations or frequency while there still is time to collect data that better meet the needs of the program.

5.1. BENEFITS OF A QUALITY ASSURANCE PLAN

A QAP must be a "living" document that is prepared in close association with the technical plans for conducting the sampling program. The QA Plan must also remain current and compatible with the devised technical program. There is no intrinsic value to the staff of a sampling program of a QAP that sits on a shelf unutilized. There are, however, significant benefits to be gained in preparing and implementing a QA Plan.

Preparation of a QAP by the program manager compels the identification of factors important to the program. It helps the manager not to overlook or gloss over details necessary to the successful implementation of the program. The key steps are (1) to conduct up-front planning and (2) to think about the details of the program prior to implementing the work effort.

Proper implementation of a QA Plan also provides definable benefits to the program manager, the manager's supervisor, and the manager's staff. These benefits include: clarity of expectations, a methodology for obtaining a valid set of data that meets the consumer's needs, and an adequate documentation trail. A QA Plan also will help in producing timely products that meet program objectives and a decrease in the amount of work that is lost or has to be repeated.

5.2. FORMAT OF A QUALITY ASSURANCE PLAN

A QAP is informational in character and should be perceived as a clarifying document for both management and the personnel conducting the program. It's structure and contents should be such that all of the factors important to the hydrologic program are considered. The plan should identify clearly the quality of the data needed and describe the procedures that will assure the prescribed quality will be produced and documented throughout the duration of the program. The plan should focus on the QA processes that provide the level of assurance desired by management but need not include the technical procedures required to collect the program's environmental data.

There is no single "correct" format for a QA Plan. The contents may vary in response to specific program needs. In general, however, a typical QA Plan will contain information on the following topics:

Title page	Identifies the plan.
Approval page	An important part of the plan that documents by signature the awareness and acceptance of the QA Plan by management and the individuals responsible for its implementation.
Table of contents	Provides the report structure.
Introduction	Should include a brief presentation of the program's purpose, scope, and description of the work. This section is intended to serve as a quick reference to the program and obviates the necessity of referring to another document for basic information the program.

Data quality aims	One of the most important parts of the QA Plan, this section defines and documents clearly the work objectives and the quality of data needed to meet those objectives.
Program organization & responsibilities	Should clearly identify the major participants in the program and their respective roles and responsibilities.
Data collection	The collection of QC data must be planned in concert with the data collection program. This section should focus on the QC data and how it relates to the environmental data collection program. It should discuss QC sample types, site selection, frequency, and sampling procedures.
Data evaluation	Documents how the QC data to be collected will be evaluated and used to ensure the technical aspects of the program are appropriate and complete.
Data archiving	Documents how all of the data, environmental and QC, to be collected will be stored, develops and documents the steps in this process.
Systems & performance reviews	Documents the review process to be implemented, including both management and technical reviews.
Corrective actions	Documents what constitutes a problem, what actions are to be taken when problems are identified, and who is to take that action.
QA reports to mgt	Documents the plans to report periodically to management on the progress of the program.

5.3. GUIDELINES FOR WRITING A QA PLAN

The management in a particular unit is responsible for ensuring that the work being done in that unit is performed appropriately. Part of that responsibility is the development and implementation of any QA Plans that are thought to be needed. For some plans, the unit chief may prepare the plan personally, while for others the chief may delegate the responsibility for preparation to other staff who are closer to the day-to-day aspects of the work to be performed. Guidelines for writing several different types of QA Plans have been prepared by the U.S. Geological Survey [1] [4] [5] [6] [7]. There are many other commercially available publications that provide guidance for writing QA Plans, but most of them are focused on industry rather than the natural sciences and require a significant amount of adaptation to be applicable.

A QA Plan for an environmental program should be prepared by the program manager. Using the format described above, here is one illustration of how a QAP is constructed:

- The program manager should prepare an annotated list of the major underlying assumption, components, and decisions for each section of the Plan.
- If the program is large enough to have several staff members, the major members of the staff should have input into the development of the Plan by reviewing it for completeness and logic.
- The program manager should expand the revised annotated list to include the types and amount of QC data necessary to ensure technical accuracy of the work as well as the QA activities necessary to ensure the work is conducted as planned.
- The program manager should complete preparation of a draft of the document.
- The draft document should be reviewed by the program manager's supervisor to ensure acceptance of the concepts and activities included and any extra work required in implementing a QA program.
- The document should be prepared in final form.
- The program manager, his or her supervisor, and any other person with a major responsibility for implementing the Plan, should sign the document's Approval Page.
- A copy of the signed QA Plan should be given to all personnel working on the program.

6. Summary

To be able to share data effectively, the data available need to be of known quality, the types and quality of data need to be consistent and comparable over time, and the data need to be both available and accessible. For a distributed database system to be useful, much less effective, it is critical for the data to be of known quality. Data quality is considered *known* if the precision and bias of the data are known, which would be the case if the data have been collected following accepted and documented procedures and if the conditions and circumstances associated with the data collection and handling are described and documented.

Common problems associated with historic databases are the lack of proper documentation, lack of ancillary data, and the changing ability to store documented records. For long-term monitoring programs, the level of documentation being collected and stored may be sufficient at the beginning of a program, but often is not kept current as new technology results in changes in some of the field and laboratory methodologies being used. Even if ancillary data are available, they are, typically, poorly associated with the environmental data that were collected and to which they apply.

Changes in the way data have been collected over the period of record make it difficult to interpret historic data. Despite awareness of potential problems, most long-term data collection programs will adopt new techniques as they are developed because the pressure to collect more accurate data is very high.

The single most important procedure to evaluate masses of data is to put them into some kind of a graphical or pictorial form. The purpose of these presentations is to search for obvious data outliers. Data outliers should raise a flag of caution, but these data should not be eliminated just because they do not fit the norm.

Collecting data of known quality requires the use of established standard collection and analysis methods, careful documentation of the work undertaken, and the implementation of a well prepared Quality Assurance Plan. Documentation is probably the single most important dimension of establishing knowledge the quality of collected data. Literally everything about an investigation or a data collection program needs to be carefully documented, including information on the methods and analytical tools employed, and especially about the assumptions that frame the program.

The potential problems organizations might have with knowing the quality, consistency, comparability, availability, and accessibility of data stored in various databases are problems that are addressed effectively by quality assurance (QA) and quality control (QC). The objectives of a QA Plan are to make clear the expectations of management and to help ensure that planned activities occur as scheduled. A QA program should not be considered a means to improve the quality of environmental data being collected. There is no single "correct" format for a QA Plan. The contents may vary in response to specific program needs. Implementation of a QA Plan provides definable benefits to a program manager, which include, as stated, clarity of expectations, a methodology for obtaining a valid set of data that meets the needs of users, and an adequate documentation trail.

A QA Plan for an environmental program should be prepared by the program manager with input from major staff members. All individuals with major responsibility for implementing the QA Plan should sign the Approval Page in the Plan.

Notes

1. Arvin, D.V. (1995) *A Workbook for Preparing Surface Water Quality-assurance Plans for Districts of the U.S. Geological Survey, Water Resources Division*, USGS Open File Report 94-382, 40 p.

2. Hem, J.D. (1985) *Study and Interpretation of the Chemical Characteristics of Natural Water*, U.S. Geological Survey Water Supply Paper 2254, 263 p.

3. Hirsch, R.M., Alexander, R.B., and Smith, R.A. (1991) *Selection of Methods for the Detection and Estimation of Trends in Water Quality*, Water Resources Research, Vol. 27, No. 5, pages 803-813, May 1991.

4. Knott, J.M., Glysson, G.D., Malo, B.A., and Schroder, L.J. (1993) *Quality Assurance Plan for the Collection and Processing of Sediment Data by the U.S. Geological Survey, Water Resources Division*, USGS Open File Report 92-499, 18 p.

5. Schroder, L.J. and Shampine, W.J. (1992) *Guidelines for Preparing a Quality Assurance Plan for District Offices of the U.S. Geological Survey*, USGS Open File Report 92-136, 14 p.

6. Schroder, L.J. and Shampine, W.J. (1995) *Guidelines for Preparing a Quality Assurance Plan for District Water-quality Activities of the U.S. Geological Survey*, USGS Open File Report 95-108, 12 p.

7. Shampine, W.J., Pope, L.M., and Koterba, M.T. (1992) *Integrating Quality Assurance in Project Work Plans of the U.S. Geological Survey*, USGS Open File Report 92-162, 12 p.

8. Vellman, P.F., and Hoaglin, D.C. (1981) *Applications, Basics, and Computing of Exploratory Data Analysis*, Massachusetts, Duxbury Press, 354 p.

LANDMASS AND WATER ECOSYSTEM DISTURBANCES IN RUSSIA AND ITS EUROPEAN NEIGHBORS: A COMPARISON

U. D. ANANICHEVA AND K. S. LOSEV
Institute of Scientific and Technical Information
and the Institute of Geography
Staromonetny, 29
Moscow 109017 RUSSIA

Abstract

Global environmental change is best measured by the scale of ecosystem disturbances. The greatest disturbances, as satellite images reveal, have occurred in inland ecosystems. This process, mainly the result of large-scale economic dt

Global environmental change is best measured by the scale of ecosystem disturbances. The greatest disturbances, as satellite images reveal, have occurred in inland ecosystems. This process, mainly the result of large-scale economic dics. This study examines and compares Russia's economically disturbed ecosystems with those of Europe and proposes a series of measures that are essential for the protection and sustainability of Russia's water resources and ecosystems.

Key Words

biomass, ecosystem, environment, landmass, pollution, purification, quality, Russia, sanitation, water resources

1. Introduction

Recently, it has become clear, that the true indicator of environmental change is not only increased concentrations of carbon dioxide or other "greenhouse" gases in the atmosphere, nor is it environmental pollution by toxic substances. Rather, it is, as well, disturbances in ecosystems and landscapes on the surface of our planet, disturbances that are visible everywhere and are especially obvious on a global scale when viewed in satellite space images of the Earth's surface. These images reveal that the largest changes have occurred in inland ecosystems.

Economic development derives from exploitation of the earth's ecosystems, its biomass, and its landmass. In the twentieth century this process has intensified on a

T. Naff (ed.),
Data Sharing for International Water Resource Management: Eastern Europe, Russia and the CIS, 61–67.

global scale to a degree unprecedented in human history, and now threatens environmental destabilization in Europe, China, and other parts of the world.

Russia is not only a land bridge between Europe and Asia, but given its water resources and landmass, it is a key contributor to ecodynamics on both regional and global scales. Absenting Antarctica and other territories that are permanently covered by frost, Russia represents 32% of the Eurasia's land surface.

From the beginning to the middle of the 1980s, using up to 4000 km^2 as the minimum unit measured, the ratio between territories that have been economically disturbed and those undisturbed, was calculated. These ratios were then recalculated in the early 1990s with smaller minimum units of up to 1000 km^2 and using three descriptive categories: undisturbed, partially disturbed, and completely disturbed. [6] The results revealed that Russia contains the largest mass of territory in the northern hemisphere that is either only slightly or entirely free of large-scale economic development.

On its northern, eastern, and southern borders, Russia is isolated from other states by water, desert, and extended topographical barriers, but no such isolating impediments exist on its western (European) boundaries. Therefore, this study will compare, from an ecological viewpoint, the use of landmass in Russia with uses of land in selected European countries.

2. The Degree of Ecosystem Disturbance Caused by Development

Russia is surrounded in the west by bio-geographical regions with destabilized ecosystems. It is also evident from the data, that in Western Europe, the only ecosystems to survive more or less intact were those in Scandinavian countries where there was a low density of anthropogenic loads on the natural systems. However, it is necessary to bear in mind that there are many nuclear power stations in Scandinavia; Sweden, for example, has twelve and Finland four.

In Sweden, nuclear power stations produce up to 50% of the national electrical energy supply, while in Russia only 12% of electrical energy derives from nuclear power. [2] The partially unaffected ecosystems in neighboring Belarus and the Baltic countries play a positive environmental role for those nations and for Russia. However, owing to the Chernobyl atomic plant accident, Belorussian Polesie, a national forest in Belarus, is contaminated by high concentrations of radiation.

As indicated, there are huge tracts of Russian land in which ecosystems remain undisturbed by economic activities and have therefore managed to survive intact. Among such regions is the Eastern Siberia taiga, which covers a total area of 6077 km^2 and includes Lake Baikal and the Kamchatka Peninsula. Another such area, constituting significant untouched virgin forests and including wetlands, was incorporated into a province of western Eurasian tiaga whose total area — but principally within Western Siberia and a European pan handle of Russia — is three million square kilometers. Finally, the High Arctic and the Southern tundra have almost completely survived. They occupy about 2.8 million km^2 within Russia.

These data require a new reckoning of Russian territory ecologically unaffected by economic development. It is now possible to state with considerable accuracy that such regions constitute not 40% to 50% of Russia's landmass, but between 60% and

65%. This estimate of unspoiled ecosystems in Russia conforms with the data available for biomass reserves.

Judging by the maps of biomass reserves of modern and restored vegetation cover reported in the Russian government's recent report on the country's environmental condition, the most ecologically disturbed regions — those with a loss of biomass greater than 50% — are the Upper Volga, the North Caucasus, the area of the Ob River steppes, and the Ussuriysk - Hankayskaya Lowlands. [1] Generally, in other densely populated areas the parameter of disturbance is not less than 40%. In the northern and the northeastern regions of Russia — in the tundra, the forested tundra, and taiga ecosystems, where the loss of biomass is only 15% — significantly more favorable conditions prevail. [7]

Changes in the Earth's surface resulting from the activities of humankind not only alter the planet's ecosystems, they can seriously disturb basic ecological functions by unbalancing nature's biochemical order which governs the stability of the environment everywhere. Such changes have been observed over a long period. They are apparent in the rapid reduction of ecologically important areas as human settlements spread across the globe and sedentary populations in many areas grow at an unsustainable rate in relation to the environment, in the exponential rise of economic development that consumes energy and resources, in the necessity of increased irrigated agriculture demanded by increasing population growth rates, and in the ubiquitous air and water pollution that accompanies these developments.

We may now examine one of the most important components of Russia's inland ecology: its water ecosystems.

3. Water Pollution

Russia is well supplied with water resources. The natural average annual discharge rate is 4300 km^2 for all uses. However, only an insignificant portion of both surface and ground water supplies are utilized — less than 2.5% of the mean, multi-annual surface discharge and only 5% of ground water resources are withdrawn. In 1995, the entire volume of water withdrawn was about 96.9 km^2 and in 1996 96.6 km^2. Russia receives an average of about 5% of its water supply from its immediate neighbors, namely, Kazakhstan, Mongolia, and Ukraine, while water flowing out of Russia into countries contiguous to its borders — chiefly Belarus, the Baltic nations, and Ukraine — exceeds the inflow by a factor of about 1.5, resulting in a small marginal net loss. Seventeen percent of Russia's water is used for irrigated agriculture and ancillary agricultural purposes, 17.5% for drinking, and 54% for industry (in 1996 ground water withdrawals amounted to only 17% of the total supplies used). [1]

In comparison to Russia, many European states use more than 50% of their national water supplies. For example, Belgium utilizes almost 100%, Bulgaria, 65%, Germany 50%, and Ukraine 56%. Only Switzerland, Sweden and Norway resemble Russian in that they use quantities lower than 50%. In regard to total water withdrawals on a per capita basis, Russia falls somewhere in the middle of the annual averages among the European states (about 700 km^2),

4. Russia's Wastewater

In 1995, the volume of wastewater in Russia amounted to 62.1 km^2 of which 24.4 km^2 was considered to be highly polluted. These figures have not varied since 1989. However, the volume of treated water has been decreasing; only 2.3 km^2 was treated in 1994. In the same year, the volume of drainage waters polluted with such toxic chemicals as nitrogen and phosphorus combinations was estimated to be 6.8 km^2. [1] Other drainage waters are officially considered to be conditionally clean, though they actually require 50 to 100 times more purification. There are large volumes of untreated and poorly treated waste water in Russia, a condition which distinguishes Russia from Europe where, in most countries, up to 100% of the waste water is totally or partially cleared of biological and other pollutants.

The quality of drinking water in Russia is low: 20.4% of drinking water supplies tested in 1993 did not satisfy official sanitation and chemical standards. For example, 11.2% of the test samples failed the microbiological organisms standards, and of those samples 4.3% constituted a real threat to health because they exceeded by 20 times the acceptable limits of bacterial content. It was largely for this reason that in 1993, twenty-one flare-ups of various water-borne diseases which made almost 3000 persons ill were recorded . The quality of drinking water has continued to decline. In 1980, 15% of test samples failed to meet purity standards and by 1994 the figure had risen to 30.5% [1]. The pollution of surface waters continues to increase despite efforts to control waste dumping into water resources.

Two-thirds of the water pollution occurs in the European regions of Russia. Practically all of the country's large rivers are polluted. In addition to point sources of pollution, there are other diffused sources such as agricultural run-offs, urban growth, road building, and many other infrastructure projects as well as both dry and wet pollutants carried in the atmosphere. [4]

5. Water Quality in European Countries

In Europe, despite huge investments in efforts to clean up the fresh water resources, the overall quality remains generally low. European nations (excluding those which were formally part of the Soviet Union) classify water quality into two categories labeled IY and Y in accordance with the degree of stream eutrophication. IY indicates highly polluted and Y extremely polluted, covering heavy metals and organic substances. Presently, data reveal that there is widespread water pollution in both categories throughout Europe. [6]

Agriculture is a prime source of pollution of European rivers and other water sources, contributing such contaminants as phosphates, nitrates, and pesticides. The degree of pollution discovered in various test samples taken in the late 1980s varies greatly from country to country, ranging from three percent over the allowable nitrate contents in Great Britain to 32% in Romania. On the other hand, the permissible limits of nitrates also varies among European nations. For example, in Netherlands and Sweden 11 mg/l (liter) are allowed while Finland tolerates up to 30 mg/l, Switzerland 40mg/l, and the United Kingdom 50 mg/l. Furthermore, again in the late 1980s, excessive concentrations of many pesticides were found in the ground waters of all

European states. For example, allowable standards were exceeded in Netherlands by up to 30%, Norway by up to 20%, and in 1990 tests of ground water for drinking in the UK, revealed that the amount of atrazine and simazine in these waters went beyond the established standards by 20% and 13% respectively.[6]

The waters of the continent of Europe (including European Russia), are significantly polluted. The pollution of Europe's waters is a consequence of the contamination of other parts of the environment. In Europe, particularly in Western Europe, the intensive construction of hydraulic engineering projects, the channeling of small rivers and creation of artificial rapids, and even the reshaping of coastlines has been the typical pattern. In Russia, the pattern has been otherwise. There, huge reservoirs-especially in the Volga River basin were built-and artificial streams were created, all of which sharply reduced the velocity of the current thereby stimulating eutrophication.

The experience of developed nations has shown that it is impossible to formulate sound policies that protect and preserve water resources without taking into account public opinion. Scientists and other specialists long ago issued warnings about the catastrophic effects that several large-scale water projects, particularly large dams and river diversions, could have. The Great Lakes Region Project in the US and the Thames River Project in the UK were undertaken with public opinion behind them only when the ecological situation became critical and citizens became concerned.

6. Strategy for the Development of Russia's Water System

It is widely accepted that preventative maintenance is the most effective policy. But such a policy cannot be implemented without a goal-oriented national government which mandates its authorities to conduct their work on the basis of this principle. The development of an effective long-range strategy for the protection of Russia's water resources requires two conditions:

- Those territories that are as yet undisturbed by large-scale economic development, as described in the first part of this study, must be, as far as is possible, maintained in their natural condition to ensure the survival and sustainability of the biosphere. These territories with their abundant water resources constitute Russia's ecological and environmental capital for the future.

- In the opinion of many Russian specialists, water quality is a complicated function of pollution related to the anthropogenic load of the source.[5] Thus, the constant assessment of the vulnerability of water resources to catastrophic anthropogenic loads together with the institution of practical measures for protecting and maintaining water quality is essential.

7. Required Measures for the Rational Use and Protection of Water Resources

Certain key measures must be taken to ensure that Russia's water resources would be utilized rationally and protected for the future.

- The development of a comprehensive network of testing and observation stations whose function will be to monitor the water regime and its quality, employing such methods as remote sensing and satellite images.
- The establishment of good, uniform standards for collecting and analyzing water data, for record keeping, and for archiving.
- The establishment of firm rules for ecologically-admissible water withdrawals.
- The establishment and collection of payments for special water uses, including the disposal of polluted and thermal water.
- The re-assessment of unexploited and exploited ground water reserves, the placement of ecologically determined restrictions on ground water withdrawals, and the monitoring of underground waters as a part of the general system for monitoring and testing surface waters all over Russia.
- Taking measures for making ground water the chief source of drinking water in Russia.

References

1. Government of Russia (1996) *Gosudarstvenny Doclad o sostoyanii okruzhauschey sredy Rossii in 1995 godu* (The State Report on the Condition of the Environment in the Russian Federation in 1995), Moscow, ECOS - inform, 452 pp.

2. Hannah L., Lohse D., Hutchinson Ch., Carr J.L., Lankerani A. (1994) A Preliminary Inventory of Human Disturbance of World Ecosystems, *Ambio, Nos. 4-5.* 246 - 251.

3. Kluev N.H. (1996) *Ecological-geographical Position of Russia and its Regions.* Moscow Institute of Geography, 161 pp.

4. Losev K.S., Gorshtov V.G., Kondratiev K.Ya, el al. (1993) *Problemy ekologii Rossii.* (Ecological Problems of Russia), Moscow VINITI, 351 pp.

5. *Natsionalny Doclad. Strategicheskie resursy Rossii* (1996) (National Report of Strategic Resources of Russia), Informational Materials, Moscow. 122 pp.

6. *The World Environment* (1992), Chapman & Hall, London, 884 pp.

7. World Resources Institute (1989) *World Resources 1988 - 1989*. Basic Books, New York, 372 pp.

DATA COLLECTION FOR THE IMPROVEMENT OF ENVIRONMENTAL IMPACT ASSESSMENT (EIA) PROCEDURES AND OF HYDRO-TECHNICAL UNITS IN THE UKRAINE

L. YA. ANISCHENKO, Cand. Tech. Sc.
Ukraine Scientific Research Institute of Ecological Problems
6 Bakulina Street
310166 Khariv
UKRAINE

Abstract

This article presents an outline of general Environmental Impact Assessment (EIA) principles in relation to Ukrainian water resources, using two extensive case studies. The author describes the EIA of large hydrological schemes in the Ukraine together with methods for conducting environmental impact assessments of hydro-technical projects. It is proposed that EIA procedures be used for water quality control. The compilation of data will enable the solution of analogous problems of water resources elsewhere.

Key Words

atomic energy, data, ecology, ecosystem, EIA, HPS, hydro-technology, reservoir, Ukraine

1. Introduction

For some years at USRIEP (Ukraine Scientific Research Institute of Ecological Problems), a unit of the Laboratory of the Ecological Engineering of Water Systems and Environmental Impact Assessment (EIA), a particular assessment methodology has been elaborated. This is an EIA method applied to economic development in industrial (mainly infrastructure projects), municipal, and agricultural sectors, in both the planning stages and those already implemented. In the process of developing the methodology, specific characteristics of regional water systems and corresponding water protection strategies were examined. The units of the study encompassed entire river basins and their individual components: canals, normal reservoirs, water-cooling reservoirs, and regulated stretches of the river.

In the domain of the EIA, integrated hydrological projects, and investigations were conducted on four main issues:

T. Naff (ed.),
Data Sharing for International Water Resource Management: Eastern Europe, Russia and the CIS, 69–83.

- Integrated solutions to problems of regional water engineering and water protection systems
- Investigation of the functional interconnections in the hydrological system.
- The development of multilateral measures for the protection of water resources and the evaluation of the ecological consequences of their implementation
- Formulation of normative standards for the preservation of water resources and for EIA

The evaluation of the hydro-ecological impact of integrated water projects on fresh water bodies was performed in cooperation with the Institute of Hydrobiology of the National Academy of Sciences of Ukraine. Included in the investigation were the ecological mechanism of water quality and detection of the functional links between the abiotic characteristics of the aquatic ecosystem (water surface area, stream depth, bank slope angle, rate of flow, water delivery regime, temperature, illuminance), the biotic associations (phytoplankton, bacterioplankton, microbenthos, higher aquatic plants), and the water quality indices (organic content, gas regime, pH, biogenic concentrations heavy metals content).

Experiments with hydro-ecological evaluations were performed at a research site situated on the Dnipro-Donbas canal. During the construction of the canal, the station was equipped for field investigations of an experimental water protection system of the canal. The tenth pumping station along the canal was selected as the research site because it allowed for a suitable range of ecological and geographic factors including the fact that the locale is very close to the Dnipro-Donbas canal, the Orelka river and Orelka reservoir. Moreover, the Dnipro-Don rivers watershed boundary is situated 20 km from the station and Krasnopavlovskoe reservoir is 40 km from the station. This reservoir is used to supply water for the city of Kharkiv and the Dnipro and Severski Donets rivers are situated about 100 km distance.

At the experimental site there were provided for experiments a fully equipped laboratory, closed cycle water shoots, and a basin. Scientific experimental investigations included studies of background water, characteristics of water borne biota and streams along the canal route, and the elaboration of water hydrological protective measures.

The compilation of experimental data allowed for the priority ranking of various methodological and scientific problems which could be solved at the research site, as well as the validation of methods whose purpose was accurate evaluation of the impact of various types of water use on surface water quality, and ways to control intrabasin processes.

Data obtained on functional hydro-ecological relationships enabled the development of a mathematical model and the generalization of a methodology for the ecological assessment of the impact of hydrological projects on the quality of natural waters. The model demonstrates the relationship of three functional indices: BODult, DO content, and phytoplankton biomass (B) as a function of the morphological characteristics of the water body. The model is based on the equation of the balance of productive-destructive processes caused by aquatic biota in subsystems of plankton, benthos, and higher aquatic plants [1].

In the last few years at the experimental site investigations aimed at determining the coefficients of the biogenic and toxicogenic limitations of production processes on stream ecosystems were undertaken along with studies of the patterns of distribution of the main components of these ecosystems in relation to heavy fluctuations of water quality indices. The results obtained from these researches will strengthen the quantifiable bases of hydroecological evaluations and will expand the application of the mathematical model.

Software was developed which allowed any experienced ecologist to perform polyvariant computations of the model on a PC. The software enabled the input data with line-by-line help to be presented in tabular form [2].

2. EIA Methodology

The main principles of EIA methodology for industries were developed on the basis of a complex systematized approach, which had been used earlier for the ecological verification of regional water engineering systems. The results of experimental and computational investigations of the ecological consequences of various types of activities were also used for this purpose. EIA is a relatively new direction of scientific research in Ukraine, and it lacks a standardized database. The general principles developed in the laboratory were systematized and included in a guideline document (A.2.2.-1-95) entitled "The Nature and Content of Data for EIA Developed at the Stage of Planning and Designing Industries, Buildings, and Other Structures: The Main Provisions for Design."

This document outlines the general principles and requirements of EIA data for buildings and structures at the pre-design and design stages. The same requirements are applicable to new constructions, the development, or reconstruction of existing industries. All state environmental protection agencies at all levels and branches and under all forms of control are obliged to adhere to the standards set forth in the guidelines document. They are also obligatory for private citizens who are involved with design and/or construction. The EIA procedures determine the ecological acceptability of designs, ensure that the environment will not be degraded and, in general, provide for ecological safety.

2.1 GENERAL PRINCIPLES

The functions of EIA are as follows:

- The determination of the prevailing environmental situation at the building sites
- The determination of possible ecologically dangerous impacts
- The determination of the severity of such impacts
- The prognosis of changes in the environment caused by construction, exploitation, and decommissioning of the designed project
- The determination of possible emergency situations
- The formulation of a set of measures for the prevention or limitation of harmful project impacts on the environment

- The establishment of rules for the compliance with the standards set by the guideline document
- The determination of the ecological and economic consequences of project implementation, and the residual impact on the environment.

Under the EIA system, it is necessary to consider a variety of options and to evaluate the impact of each option on the hydrosphere, atmosphere, flora, fauna, archeological sites, and on the technological and social environment. EIA results are used for evaluating the environmental consequences of project implementation, clearly identifying the main effects. The conclusions drawn must be published and made available to the public.

2.2 SPECIFIC FEATURES OF EIA FOR HYDRO-TECHNICAL OBJECTS

In the EIA documents the subsection dealing with the aquatic medium is considered to be the main one. It must include a detailed analysis of shifts in hydrological and hydrogeological characteristics of the water body, caused as a result of the project implementation. It is necessary to evaluate impacts on the flow, level, temperature, and oxygen regimes, on riverbed processes, siltation, content of priority pollutants in surface and ground waters (including pollutants delivered with waste water non-point discharges, and atmospheric precipitation). It is also necessary to evaluate the impact on self-purification processes, biological productivity, secondary pollution, and the destruction of organic substances.

Measurements must be made of the impact of municipal, industrial and storm waters formed at the site of the designed structure in the process of its construction and operation, with allowance for possible emergency situations. The results must reflect the distribution of assessed values indices of the area and monitoring sections, and take into account additional variable impacts. The regime and volumes of releases to tail bays of hydro-engineering units must be determined. The rationale for measures aimed at prevention and mitigation of pollutants inflow, depletion and deterioration of water resources and aquatic organisms degradation must be given.

If necessary, it is possible to include propositions for the revision of values and time limits set for the attainment of maximum allowable discharges (MAD) by industries in the zone impacted by hydro-technical construction. Variants of computations used for evaluating the consequences of project implementation on aquatic ecosystems should be based on least favorable periods, with allowance for emergency situations.

General data on the surface waters selected for a hydro-technical construction project must contain information about the catchment basin and its economic uses, and a list of monitoring points and obtained results. When assessing the impact of designed objects on surface water and aquatic biota it is necessary to take into account the following factors:

- The physical, chemical, sanitary, hygienic, toxicological, radio-ecological characteristics of bodies of water and their ecosystems

- The biological characteristics, including species composition, numbers of individual organisms, their biomass and their bioproductivity.

2.3 EIA EXAMPLES

The methods outlined above were applied in evaluating the hydro-ecological impact caused by a number of hydrological infrastructure construction projects in Ukraine, Russia, and other countries. [3] For example, the impact of the Akosautsky hydro-scheme in the Caucasus on a mountain river ecosystem was evaluated, and the effects of the Krapivinski hydro-scheme on a river of the plains were also measured. [4] Here the main task was to predict possible changes in the way the Tom River ecosystem functions which might be caused by the construction of a seasonally regulated reservoir in Russia.

The evaluation of the impact on water quality of hydrological projects still in the design stage intended for Tisa River basin were performed in the laboratory. These tests proved to be highly topical. The designs provided for the construction of partitioning dams with low head electric stations, protective dams along the river banks in the plains segment of the river, derivation canals at tributaries, and intrabasin water redistribution among tributaries. The key effects on ecosystems and water quality caused by river bed dams can be seen in the rate of flow and water depth while those for derivations and intra-basin redistributions are changes — principally in water content — at stretches below the off-take canals. A comparative analysis of various options for re-situating the infrastructure schemes on the Tisa River and its tributaries was performed which produced recommendations for taking various measures to protect the hydro-ecological system.

Work of the same type was performed for the development of recommendations aimed at improving ecological conditions in the Ingul River basin (a tributary of the Southern Bug River), for the determination of ecologically safe releases of water to the Ingulets River (in the Dnipro River basin), and to the Lower Kuban River basin in Russia.

The impact of hydrological infrastructure projects on the environment are often complex and it is therefore quite possible that the predicted effects will differ from the actual impact when such projects are completed. This circumstance stems from the absence of sufficiently rigorous methods of evaluation. Consequently, it is usually necessary to refine and revise the prognosis on the basis of further investigations during the period of project construction and implementation.

3. Methods Employed by Official Environmental Experts

When ecological impact assessments are completed, the results are presented to official ecological experts for consideration. It is understood that any project will produce some negative consequences for the environment. However, the key principle is that whatever economic advantages are gained by a project, its implementation must, in the long run, result in positive social effects. Thus, during the time that contracts are assigned and during the period of construction, it is necessary to perform field surveys of the environment in accordance with special programs that include a system of environmental monitoring.

The Ukrainian government has legislated standards and procedures for EIA reviews by panels of government environmental experts. This legislation is one of the main regulatory instruments for ensuring the ecological safety of economic activities at various levels.

The main principles of ecological security followed by panels of environmental experts and agencies of the Ukrainian Ministry of the Environment are:

- To ensure that the natural environment is safe for the life and health of the people
- To balance ecological, economic, public health, biological, and social interests, allowing for public opinion
- To provide for open, independent, objective, integrated scientific investigations, which offer alternative options and preventive measures
- To ensure ecological safety, establish the economic and regional necessity of projects and their operation
- To provide state regulation
- To ensure adherence to the law.

The application of the state's environmental expertise is intended to produce data documentation for the introduction of new methods, materials, substances, and products which would help avoid the violation of environmental norms and end practices that create a threat to human health among various operating units and agencies of production in various towns and regions.

3.1 GENERAL RULES

The official experts begin by determining the main direction of any investigation. In so doing, it is necessary to conform to the law and simultaneously to take into account the realities of each project and the site selected for its implementation.

The types of hydrological infrastructure projects that must be subjected to government environmental review are dams, dikes, water power stations, atomic power stations, water intakes, canals, and artificial bodies of water created as a result of such construction, e.g., reservoirs, waste water lagoons, filtration fields, and irrigation systems.

In the process of evaluating such activities, the experts must assess:

- The extent to which a project conforms to the requirements of the law and the recommendations of the relevant EIA
- The extent to which the design of the project conforms to the demands of environmental protection during the time of construction and implementation
- Whether there would be sufficient environmental safety in conditions of emergency.

The state's environmental experts must conduct reviews of the possible impact on the environment of hydrological projects at various stages of the investment process. The experts must investigate several kinds of activities involving both government and private investments:

- The development of schemes for the rational use of water resources, programs for water engineering systems development within one or several river basins
- The submission by investors of proposals for the development or reconstruction of hydro-technical projects in addition to previously approved schemes
- The preparation of documentation for the selection of land sites for the construction of infrastructure projects
- The proper performance of feasibility studies on plans for infrastructure projects submitted by contractors.

In some instances, government environmental experts may subject a hydrological project already in operation to an evaluation.

EIA performed at the initial stages of the investment process need not to be very detailed. The next major phase is the feasibility study at which time all EIA data are given to the authorized environmental experts. In those instances where alterations are made to approved designs by the contractor, EIAs should be revised accordingly. Proposed schemes of operation of water engineering units are required to include a preliminary general EIA with a short analysis of current environmental conditions in the area and site that will be impacted by planned activities. Plans are also required to contain a prognosis of possible changes in prevailing conditions which could be caused by the planned activities. These reports must include a comparative analysis of various schemes and options for environmental protection.

At the feasibility study phase of a project, managers must provide the government's environmental panel of experts with the following documentation for evaluation:

- A project summary including proposed designs and explanatory notes
- An EIA of the planned construction for the environment involved
- A statement concerning the ecological consequences of the project's implementation.

If the government experts feel the need, they have the right to demand any further documents, including computations, surveys, etc. that are retained at the firm which has designed the project. The documentation submitted for evaluation should include a list of all participating firms and any local agencies together with the signed agreements among them. The materials presented for EIA should include copies of all previously obtained technical specifications and conclusions of any other firms involved.

3.2 EXCEPTIONAL SITUATIONS

There are many exceptional situations that affect the application of the environmental safety rules and the work of government experts, such as the transfer of ownership of an enterprise which results in the loss of some aspects of environmental monitoring because the legal circumstances change with the ownership. Today's "Green Movement" whose original purpose was to lobby for the protection of the environment, often unites with certain other political movements which has made the "Green Movement" highly unstable. Moreover, in Ukraine, difficulties arise when an EIA is proposed for the reconstruction or augmentation of currently operating hydro-power units. Often during the long periods when such units were originally constructed and operated, other laws obtained and subsequent legislation introduced new rules for protecting the environment creating inconsistencies; at the same time, economic priorities may have changed along with environmental conditions which were often aggravated. In such circumstances, the prevailing rules for EIA and environmental safety are frequently by-passed.

Two instructive case studies may be profitably studied: The Zaporizhzha Atomic Power Station and the Tashlyk Hydro-accumulating Power Station (HPS) in Ukraine:

3.2.1. *The Zaporizhzha Atomic Power Station*

The problem of enforcing current EIA requirements were encountered when the construction of the sixth block was commissioned for the Azporizhzha Atomic Power Station. [2]

This particular power station site was selected when the decision-making processes of the former USSR were still in effect. Consequently, the political, economic, legal, scientific, and technical criteria that were applied in the selection reflected that system. At the start of construction, the region had a strong basic infrastructure, i.e., skilled manpower, transportation, a township and other advantages. The Kakhovske reservoir provided a sufficient water supply. All these factors led the authorities to increase the station's capacity to 6000 MW. The project was developed in 1985 when five station blocks were put in operation and the proposed sixth block was close to being commissioned. The ecological assessment service informed the authorities that the Zaporizka regional power station, situated about 25 kilometers from the Zaporizhzha Atomic Power Station was the main source of environmental pollution in the area.

When an EIA of the sixth block was conducted as part of the commissioning process, it was necessary to take into consideration the entire powerful energy providing complex in the region. That included the atomic and conventional power stations and the possible impact of on-going operations and the proposed new construction on the Kakhovske Reservoir fauna and flora, the surrounding atmosphere and groundwaters. When a project is essentially one of enlargement of an already existing operation, it is impossible for an EIA to influence such key factors as the conditions at the proposed construction site, the type of the cooling system and reactors, the positioning and lay-out of auxiliary units, etc.

As a result, the environmental experts involved concentrated mainly on the effects of the additional load on the environment that would be caused by the sixth block, and on measures aimed at mitigating the impact of the atomic and conventional power stations on the environment. They outlined the directions of investigations and monitoring that would have to be undertaken which would provide feedback on results of the plant operations that would allow for mitigating negative conditions.

Special attention was paid to the cooling pond. The experts investigated the processes leading to increases of salt content in the pond and the blooming of phytoplankton and bacterioplankton caused by the discharge of organic-rich waters. They examined processes of heavy metals sorption on finely dispersed particles and in the processes of radionuclides sorption and piling-up in bottom sediments, and they evaluated the real content of heavy metals discharged with purging water into the Kakhovske Reservoir. They also predicted the piling-up of heavy metals in the cooling pond bottom sediments during the time of operation. Special attention was also paid to forecasting radioactive pollution of aquatic ecosystems in water sources used for potable purposes and to the content of radionuclides in atmospheric precipitation and in land ecosystems.

Investigations performed on bodies of water in the area around the power complex indicated that the atomic power station had no influence on the thermal pollution of the area adjacent to the Kakhovske Reservoir. Such pollution was caused by the discharge of heated waters from the Zaporiska thermal power station.

3.2.2. *The Tashlyk Hydroaccumulating Power Station (HPS) in the Ukraine*

The difficulties of carrying out the official responsibilities of government EIA experts when hydro-power facilities are involved is exemplified by the case of the Tashlyk Hydroaccumulating Power Station. The difficulties stem from the close connections of political, economic, and ecological problems which have tended to be acute during the period of transition from the communist economic and bureaucratic systems.

Shortages of fuel and power resources in Ukraine have created problems of electrical energy shortages. The reliability of electric currents has declined, the energy grid of Ukraine is close to the limit of its carrying capacity, and there is insufficient power for regulating loads at peak periods. Ukraine developed and is presently undertaking a program of HPS construction whose purpose is to smooth irregularities in electrical use, regulate the AC frequency, and to provide an operational reserve of electrical power. The ratio of day and night tariffs for electricity usage is about 1:10.

The Tashlyk HPS is situated on the Southern Bug River and is a part of the South Ukrainian Power Complex which includes the South Ukrainian Atomic Power Station. This HPS is intended for supplemental power supply during periods of peak loads and for improvement of the reliability of operations in the southwestern region of the joint power grid of Ukraine.

The HPS construction was begun in 1981. The station includes the up-stream reservoir, which is a part of the southern Ukrainian atomic power station cooling pond, the HPS buildings, the downstream Alexandrovske Reservoir and the hydropower station.

Since the disaster at the Chernobyl atomic power station, the public pays close attention to problems of atomic and hydropower. In 1989 the government conducted an environmental review of the southern Ukrainian energy complex. As a result it was decided to use only three hydro-aggregates at the Tashlyk HPS instead of ten. The redundant seven had to be mothballed. Improvements to the project were also recommended. In 1991 a moratorium was placed on construction of hydro-and-atomic power complexes, but by May 1, 1991 the construction of the start-up HPS complex was about 90 per cent compete.

In 1992 twenty academic and other scientific institutes prepared and submitted to the state environmental review board proposed innovative revisions of the project. The project was provided with a complex program on social and economic development of adjacent regions. Under the new plan, it was proposed that the HPS up-stream reservoir be formed by damming up the head area of the Tashlyk Reservoir which was being used as a cooling pond for the South Ukraine atomic power station. The existing Alexandrovske Reservoir, with an 8.0 meter flood control storage level (FCSL), was to be used as the HPS downstream reservoir. The project provided for the elevation of FCSL to 16 meters for the operation of three of the aggregates without an additional water pool, 16.9 meters for the recommissioning of aggregates 4, 5, and 6 without an additional water pool, and 20.7 meters for the commissioning of aggregates 3, 4, and 6 with an additional water pool to be used for supplying the Mykolaivska oblast (administrative territory). The recommendation to create an additional water pool was motivated by a regional water shortage which was related to a stressed water balance and to deteriorating water quality in the Southern Bug River.

In response to the proposed revised HPS plan, the government environmental experts made the following recommendations:

- Of a total of 2409 hectares (ha) of land available for the project, 1022 ha would be flooded, 604 ha would be used for buildings, and 650 ha would be used for planting forests and other environmental protection measures. Because the reservoir would be situated in a deep canyon, a minimum amount of land would be flooded

- The backwater zone of the Alexandrovske Reservoir would in part be extended to the territory of Konstantinovsky, to part of the state sanctuary of "Granitno-Stepnoe Pobuzhe," also to the nature sanctuary area "Bugski," and to some land which contains rare and threatened plants will be flooded

- The Alexandrovske Reservoir would have an exceptionally high flow-through rate which would result in rather intensive turbulence in the upper layers of water in the exchange with deeper layers. Stagnation phenomena would be absent, DO content would be high at all levels so that the intensity of blue-green alga development would be low

- The production of HPS electric power in peak periods would be attained without the discharge and effluence of harmful substances

- Various engineering solutions included in the project would reduce harmful effects of the HPS on environment:

    - A fertile layer of soil would be removed from the zone of flooding

- Low fertility lands would receive fertile soil
- The reservoir banks would be diked
- At the Alexandrovske reservoir special sprinkling units for DO content augmentation in low-flow periods would be used on water discharged to the aft bay
- In dry spells, it would be possible to release flashes of water from the Alexandrovske Reservoir
- Waste water treatment units would be constructed in the sanitary zone around the project site
- Forest shelter belts and other anti-erosion and water protection measures would be undertaken
- The "Bugski" nature sanctuary would be protected against flooding
- Rare and threatened plants would be removed from the flood zone and re-planted elsewhere
- A network of entomological micro-sanctuaries would be created
- A fish-way canal and a fish nursery together with an artificial spawning area would be constructed
- All territory designated for flooding would be archeologically investigated prior to flooding
- Field observations would be performed in accordance with the rules of environmental monitoring

• The implementation of the project would produce positive social results: energy produced by HPS would be used mainly for social needs, electricity would be supplied to towns and rural regions reaching a population 19,0000 people, and the project would provide for the creation of schools and nurseries and it will help in construction of roads, centralized water supply, etc.

• During the period of HPS construction it would be necessary to conduct field surveys of the state of environment including studies of productive-destructive processes, and of self-purification and of the self-regulation capacities of artificial ecosystems

• Control over operation of the Alexandrovske and Pribugske Reservoirs would be transferred to the local authorities whose most important responsibility would be protection of the environment.

While the social and economic programs for the Tashlyk HPS project are persuasively outlined, the ecological and water engineering parts are somewhat lame and have come under continuous criticism. However, at this stage when the complex is in an advanced state of planning and soon to commence, it would be difficult to introduce new variations. An analysis of the project has revealed that to a certain extent it does deal with some of the shortcomings of earlier economic activities, ameliorates some of the harm to the ecosystem, and there are sections dealing with social issues, transportation, water supply, recreation, and soil recultivation problems.

The current Alexandrovske Reservoir is at the lowest stage of the Southern-Bug cascade of reservoirs. Above it on the Southern Bug River and its tributaries there are 192 other reservoirs and about 7000 ponds constituting a total aquatic area 740 km^2 and containing a volume 1500 mcm (million cubic meters) of water. In the regions around the Tashlyk power station there are another six rivers, 108 ponds, and ten reservoirs. Seasonal, weekly, or diurnal regulatory discharges are performed at a majority of the reservoirs and ponds.

These data indicate the necessity for a general, integrated hydrological infrastructure development plan for the Southern Bug River basin as a whole. Such a comprehensive plan would have to be based on a comprehensive environmental evaluation of the current state of the basin's water resources, an assessment of the man-made water units, a system of environmental monitoring, and environmentally linked activities for the cascade of reservoirs involved. Only on the basis of such comprehensive planning and discussions with local authorities will it be possible to solve the urgent environmental problems of the Alexandrovske Reservoir capacity augmentation scheme.

The situation is aggravated by a group of people who actively fight against completion of the project. They say that the project will damage the regional landscape of the "Granitno-Stepove Pobuzhe" sanctuary park. There have been a number of public hearings and discussions with the planners of the project. Presently, the environmental review board continues the process of evaluation.

The ecological review process of the Dnister power station was performed in a less controversial atmosphere. The project of the Dnister multipurpose hydraulic complex was developed in the 1960s within the framework of an integrated power grid for the former USSR. It was intended for producing additional power supply at peak hours of usage. In addition, the complex provided for flood protection and made it possible to augment the middle and lower segments of the Dnistro River so that they could be used as reliable sources of water supply for the domestic, industrial and agricultural sectors.

The project provided for the construction of three reservoirs, the largest of them being the Dnister Reservoir with a volume of 2000 mcm of water. The second, a balancing reservoir, is intended for smoothing pulsations of water releases from the upper Dnister Reservoir. The third, an HPS reservoir, is situated above the balancing reservoir; the normal water level drop from the HPS reservoir to the balancing reservoir is 150 meters (m). The HPS reservoir is intended for the use of surplus energy in the grid during hours of low usage and in hours of high usage water from the HPS reservoir can be delivered through pumping-turbine units to the balancing reservoir to produce additional power. The HPS consists of the following main structures: the upper basin, the pressure boosting unit, and the lower balancing reservoir with protective structures.

The construction site of the HPS is situated on the right bank of the Dnister river; eight km below the Dnister power station dam, is the Sokiryany region of the Chernovtsy oblast. The normal water level of the basin is 229.5 m with volume 38.8 mcm. The site of the lower balancing reservoir is situated in the Chernovtsy and the Vinitska oblasts of Ukraine and Moldova — its normal water level is 77.1 m and its volume 61.2 cubic meters.

The operation of the Dnister hydraulic complex since 1974 has caused radical shifts in the natural characteristics of the river flow. Seasonal fluctuations of the flow have been considerably reduced. Because the turbines are operated two or three times a day, a new fluctuation pattern of the Dnister's diurnal flow fluctuation has resulted while construction of the balancing dam continues.

As water from the reservoir is discharged to the tail bay at a depth below the reservoir thermocline, the temperature regime and DO content are distorted from the Dnister segment of the balancing reservoir to the Dubosary Reservoir. As suspended solids inflowing from the head segments of the river are sedimenting in the Dnistro Reservoir, water turbidity below the balancing dam is reduced.

Such changes in the Dnistro River characteristics have caused serious damage to ecosystem. Fish migration routes were physically cut by the power plant, and the smoothing of spring freshet peaks damaged valuable spawning areas situated all along the river from the dam down to the Dnistro firth (inlet) near the city of Odessa. The thermal regime in the middle stretch of the river does not correspond to the natural conditions necessary for spawning. Reduced DO concentrations below the power station negatively impact the ecosystem, and reduced turbidity of the river water has resulted in changes in the illumination regime at deeper levels of the water, causing an increase of macrophytes growth.

Thus, the power station construction has negatively impacted some of the natural characteristics of the Dnistro river, such as river flow distribution, levels of water temperature and turbidity, and DO content. The main effects are caused by hydrotechnical construction in the upper basin and at the balancing reservoir, and by waters accumulated in the reservoirs.

The total volume of the upper and the lower reservoirs of the HPS amounts to 108.9 mcm. At normal pool levels, that is 2.3% of the 30Q20 annual discharge which in the balancing reservoir section amounts to 4.76 km^3 which reduces annual discharge by 0.05%. Filtration losses from the reservoirs amounts to 0.8 mcm/yr, but they are replenishable for the Dnistro river. HPS structures apparently affect the volume and annual distribution of the river flow only marginally. Fluctuations in the water level of the balancing reservoir caused by daily discharges of peak flows can be reduced by strict observation of the dam operation regulations.

The inclusion of the HPS in the Dnistro hydropower station system will allow an increase of the residence time of the water in the segment between the Dnistro Reservoir dam and the balancing reservoir dam. This will lead to a more intensive thermal exchange with the atmosphere and to a slight reduction in the river water temperature. The increase in the residence time and water mixing in the turbines will lead to higher DO content in the water of the balancing reservoir; the water discharged from this reservoir will have a DO level approaching that of the natural river water. Turbulence arising during water flow through the HPS turbines will lead to more intensive erosion processes in the balancing reservoir and to a decrease in sedimentation, and, as a consequence, to higher turbidity in the tail bay of the reservoir.

The present day water quality in the zone affected by the balancing reservoir was measured between 1993 and 1995. The following indices were used. total amounts of salt ions, chlorides, sulfates, ammonium, nitrite and nitrate nitrogen, phosphates, and BOD_5, DO, COD. The majority of these indices were within normal limits, with the exception of ammonium nitrogen (its content amounted to 13 times more than normal values), nitrites (up to 25 times above normal), and BOD_5 (which was sometimes up to 1.2 times above normal). Water turbidity in the upper part of the balancing reservoir amounted to 6 g/m^3, and in the tail bay 9 g/m^3. Among toxicants the most important were organochlorine pesticides, copper, zinc, and phenols. In the Dubosary Reservoir, situated on the lower Dnistro River, iron, fluorine, manganese, lead, and nickel were detected in addition to the pollutants listed above.

Human impact on water quality is increasing from the upper stem of the Dnistro River basin to the lower stretches, the result of municipal waste waters discharges by the towns of Drogobych and Boryslav as well as agricultural waste waters and other non-point discharges. Sugar factories also contribute much water pollution in the affected area. The Dnistro Reservoir serves as a kind of buffer against toxicants discharged into upper segments of the river but waste waters of stock forms contribute dangerous pollutants to water resources in the region. Moreover, many municipal, industrial, and other economic units discharge their wastewaters into the rivers without sufficient treatment.

Most municipal treatment stations function at an efficiency rate no higher than 50% to 70%. Large quantities of contaminants come from non-point discharges into tributaries, which are consequently significantly polluted, and from there into the Dnistro River and the balancing reservoir. Pathogenic bacteria were detected in the river and its tributaries as well as high levels (as much as 2 - 10 MAC) of BOD; such levels often leads to DO deficits. Ammonium nitrogen concentrations in the area's waters have also periodically attained the value of several MAC. Such high levels of organic pollution in the Dnistro tributaries was confirmed by the vigorous growth of phytoplankton and by the results of bacteriological analyses.

However, the Dnistro River at the HPS site has a rather high self-purification capacity. The sanitary and hygienic indices there were almost the same as in the upper stretches of the river. The impact of the HPS on the hydro-ecological conditions in the Dnistro River below the balancing reservoir was evaluated as being insignificant. Among the positive effects expected to result from the construction of the HPS are an elevation of the river's water temperature up to its natural level, an increase in the DO content because of aeration, and an invigoration of the river's self-purification capacity. The negative effects would be increased turbidity with a possible release of harmful substances from bottom sediments.

Because the public was given insufficient information during the planning stages of the HPS project, its attitude has been ambiguous. The local population is particularly concerned about the possibility of periodic flooding of some of the region's housing. But the EIA data prove that this region is free of geological or tectonic conditions that could cause the seismic failure of the project's dams owing to additional loads put on them when the upper and balancing reservoirs are filled with water. Also, special engineering methods will be applied to limit the action of filtration on the stability of the reservoir banks.

In order to protect dwellings in the vicinity of the project, protective embankments and pre-embankment drains will prevent flooding. Moreover, should there be flooding despite these precautions, plans have been drawn up for the quick evacuation of the affected population and for the removal of flood waters. The project design ensured that natural wildlife refuges and areas of aesthetic value would not be directly affected when the project becomes operational. The plans call for the creation of a new belt of forests to replace those forested areas in the project environs that would be flooded, and several fish farming units will be established to compensate for possible losses of fishing sites. Finally, the local populace would be fully informed about all of the provisions for ecological and social safety embodied in the planned project.

Because the geological, hydrological, and hydrogeological dimensions of the in the area of the HPS construction are complicated, the EIA panel insisted that prior to commencement of the project it would be necessary to develop a system of environmental monitoring for the region. In general, the EIA concluded that the Dnistro HPS project would be of considerable importance for the economy of Ukraine, and since the HPS needs no fuel, its impact on the atmosphere, water resources, land and aquatic ecosystems in the region were considered to be acceptable. The EIA report was, on the whole, considered by the authorities to be satisfactory.

4. Summary

Despite serious national economic difficulties in Ukraine, the requirements of the environmental legislation are strictly observed. Each project must undergo an EIA by environmental experts, a procedure rooted in sound legislation and regulations which include provisions for data collection, standardization, and sharing among regional environmental centers, and the further development of EIA methods and comprehensive indices for making evaluations. Many Ukrainian planning and design institutes are developing EIA procedures which are applicable to various regions.

5. Notes

1. Romanenko V.D. et al. (1990) *The Ecological Evaluation of the Impact of Hydro-technical Construction on Water Resources* , Naukova Dumka, Kiev.

2. Anischenko L.Ya. (1996) Problems of Environment Protection. Perspectives on EIA Methodology and the Results of the Development of Ecological Expertise, *Proceedings of UkrNTSOV Institute 1991-1996*, Kharkov.

3. Anischenko L.Ya. (1989) Water Quality Control in Regional Water Engineering Systems, *Upravlenie vodnymi resursami sushi. Teoriya i praktika*, Nauka, Moscow.

4. Anischenko L.Ya. and Churaevskaya N.N. (1991) Evaluation of the EIA of the Aksautsky Hydro-technical Complex Design: Methodological and Practical Aspects, *Proceedings of the First UNESCO Workshop on EIA: Methodological and Practical Aspects*, Moscow.

5. Anischenko L.Ya. and Sverdlov, B.S. (1991) The Ecological Assessment (EIA) of the Krapivinsky Hydro-engineering complex on the Ecosystem of a Plains River, *Proceedings of the First UNESCO Workshop on EIA. Methodological and Practical Aspects* , Moscow.

PROBLEMS OF DATA:

Monitoring the Quantity and Quality of Water Resources in Belarus

BORIS FASHCHEVSKY
University of Modern Knowledge
Kalinovskogo Strett. 48-49
220086 Minsk BELARUS

Abstract

Issues of water resources monitoring, record keeping, and databases in Belarus are examined. In this context, the quantitative and qualitative characteristics of the country's water resources are also analyzed and the state of national water resources are described.

Key Words

atomic energy, Belarus, degradation, ground water, hydro-chemical, lakes, monitoring, pollution, quality, rivers, surface water, thermal

1. The Hydro-Geographic Characteristics of Belarus

The Republic of Belarus occupies approximately 208,000 square kilometers in the center of the European Continent. Belarus borders Poland in the west, Lithuania in the northwest, Latvia in the north, Russia in the northeast and east, and Ukraine in the south. The Belarus terrain is, for the most part, flat, and its altitude above sea level varies between 80 and 750 meters.

The climate is moderately continental. Average annual air temperature drops from the southwest to northeast, varying from 7.40 C to 4.40 C. The average annual precipitation varies from 520 mm in the South to 720 mm in the North. The country has a well-developed hydrological profile consisting of a network of numerous rivers, lakes, reservoirs and channels. The average density of the river networks is 0.45 km/km^2 and the total number of rivers is 21,000, with 1,900 of them being more than 10 km in length. There are seven large rivers, that is, more than 500 km long: the Berezina, Viliya, Neman, Sozh, Pripyat, Zapadnaya Dvina and Dnieper. The total annual accumulated runoff in the territory of Belarus is 35.3 km^3; if the inflow from bordering is taken into account, the total runoff increases to 56.0 km^3.

T. Naff (ed.),
Data Sharing for International Water Resource Management: Eastern Europe, Russia and the CIS, 85–94.

The hydrological network includes about 10,000 lakes, 470 of them having an area of more than 0.5 km^2. The largest lake is Naroch, 79.6 km^2 in area, the deepest is Dolgoye, whose size is 53.6 km^3. The lakes are widely used for the development of fish breeding, as a source of water supply for the surrounding population, and as cooling ponds for atomic and thermal power stations — namely, Drysvyaty, Lukomlskoye, and Beloye. The major hydro-geographic division runs from southwest to northeast and separates rivers into Baltic and Black Sea slopes. Within the territory of Belarus there are about 70 reservoirs and 500 ponds. The largest reservoir, Vileyskoye, is situated in the upper stem of the River Viliya; its area is 75 km^2 and its capacity is 260 million m^3. Large-scale drainage amelioration has been undertaken throughout Belarus, but especially large areas were drained in the basin of the river Pripyat.

It should be mentioned that in 1976, during the Soviet era, the largest water economy system of interbasin redistribution of water resources was created in Belarus. It was the Vileysko - Minskaya Water System for the capital city of Minsk which provided water for its two million inhabitants. Presently, there is a water transfer from the Viliya River to the Svisloch River. About 200-250 million cubic meters of high quality water from the Vileysko - Minskaya Water System travels 140 km straight from the northwest to the southeast. The length of the transfer canal is 62 km with a series of five pumping stations along its banks which are necessary for raising the water 75 meters along its route. The system's maximum potential water transfer capacity is 250 million m^3.

Intensive industrial and agriculture expansion has considerably changed the hydrological and hydrochemical regimes of many rivers and lakes in Belarus. Moreover, the Chernobyl atomic plant accident contaminated about 20% of the territory of Belarus with radionuclides. It has caused about 100,000 inhabitants in the contaminated region to emigrate to clean regions of the state.

2. Water Resources Monitoring and Record Keeping

Currently, there are five systems of monitoring that provide information about the quantity and quality of surface and underground water resources in Belarus:

- The Hydrometerological Service which conducts hydrological, hydrochemical, radiation and biotic observations
- The Ministry of Health which conducts bacteriological and hydrochemical observations
- The Ministry of Housing and Communal Economy conducts hydrochemical observations above and below water treatment works facilities
- The Ministry of Natural Resources and the Environment monitors water withdrawals from natural sources, water utilization and the depositing of sewage into waterbodies.
- EU "Belgeology" maintains observation of underground water regimes and their quality.

The databases for the data and information collected by these units are kept only in the Hydrometeorological Service, in the Ministry of Natural Resources and the

Environment, and in the Institute for Water Resources Management. The largest number of observations are made by the Hydro-meteorological Service on rivers, lakes, canals and reservoirs. Presently, there are 120 gauging stations that monitor rivers and canals while 14 gauging stations record data for lakes and reservoirs. There is one gauging station per 1550 km^2 in Belarus. The measurements made at these stations include:

- levels and velocity of water
- discharge
- turbidity
- discharge of suspended and bed loads
- granulated composition (varying diameters in mm) and density
- temperature
- thickness of ice and the depth of snow on ice
- ice phases in the area of gauging.

For lakes and reservoirs there are additional measurements:

- temperature of water near the bank
- temperature of water at various depths
- heat-holding capacity.

The measurement for the hydrochemical content of water bodies are made at 75 permanent points approximately once a month. They include:

- odor of the markers
- total mineralization
- transparency
- color in units
- total hardness (calcium carbonate)
- phenols & oil products
- suspended solids
- surfactants
- main ions (HCO_3, Na, Ca, Cl, SO_4, K, Mg)
- pesticides
- pH content
- ammonia (as N)
- biochemical oxygen demand (normally BOD_5)
- ammonia nitrogen (as N)
- chemical oxygen demand (COD)
- phosphates
- conductivity
- nitrate (as N)
- dissolved oxygen
- silicon
- percentage saturation of oxygen
- total phosphorus
- carbon dioxide
- heavy metals*

*copper, lead, zinc, nickel, chromium, iron, cadmium, mercury, manganese

The points at which samples are taken are usually situated above and below inhabited sites. Additional samples are taken at two points of depth from lakes and reservoirs.

After the 1986 accident at Chernobyl, permanent observations for radioactive pollution are conducted in five basic rivers: the Rechitsa stem of the Dnieper; the Gomel stem of the Sozh; the Mozyr stem of the Prypyat; and the Svetilovichy stem of the Besed. Samples for radionuclides — cesium137, strontium-90 and beta-activities — are taken every month from surface waters. The Hydrometeorological Service measures phytoplankton, phytopherifiton, zooplankton and zoobentos several times a year.

The underground water regime measurements are taken in 1564 boreholes drilled in all basic aquifers. They are united into 216 stations [4]. One hundred thirty seven stations are intended for studying the conditions within which the ground water resources have formed with a view to revealing and evaluating anthropogenic changes related to global and regional sources of pollution. Seventy-nine stations (850 boreholes) are situated at points of large water withdrawals (plants, towns, etc.) for the purpose of measuring the impact of the Chernobyl incident on the condition of the ground hydrosphere.

The measurements at these ground water regime stations include: the depth and temperature of the water and samples are taken for purposes of physical and chemical analysis. Depth and temperature measurements are taken between three and ten times monthly and samples for analysis are taken from one to four times annually. The Ministry of Health Protection takes ground water samples for bacteriological and hydrochemical analyses once a month and the Ministry of Housing and Communal Economy takes samples for analysis from one to three times a month.

Once a month, under the supervision of the Ministry of the Environment and Natural Resources, various organizations such as collective farms monitor withdrawals, the utilization of ground water, and the dumping of sewage into ground water sources.

It is necessary to emphasize that the majority of organizations and other enterprises that collect the data actually have no equipment for measuring the quantity, quality and withdrawals of water. The only information that is given is that which can be collected at power pumps and from other incidental observations. After all the surface and ground water data are collected, analyses are made and conclusions drawn.

On this basis, official yearbooks are published on the following subjects: hydrological data; hydrochemical data; water utilization; the pollution of rivers by sewage; and a general water cadaster that includes data on surface/ground water and sewage. In addition to the yearbooks, there are periodically published books based on many years of accumulated historical data — one of these books was drawn from the record of ten to twenty years of data. The yearbook of hydrological data, includes daily levels and discharges of water, lists of gauging stations with a brief explanation of their location.

The yearbook on surface water quality includes a list of points of pollution control with a short description of location and period of action, and a survey of the hydrochemical regime of rivers, canals, lakes and reservoirs. It indicates in the form of tables the standards which establish the maximum allowance for hydrochemical concentrations in water bodies.

Perhaps the most interesting and useful of these year-books is the *State Water Cadaster* [4] which covers the following general subjects:

- general characteristics of water resources utilization and quality
- the hydrographic network and changes resulting from the impact of drainage amelioration (Table 1)

- the total hydrological, hydrochemical and hydrobiological measurements
- hydrometeorological conditions, river flows and the flows of suspended sediments
- the quality of surface waters
- total hydrogeological measurements
- total ground water stocks including various changes (Table 2)
- quality of ground water sources
- utilization, consumption, and water led away
- sewage pollution of rivers (Table 4)
- long-time average flow of main rivers of Belarus (Table 5)
- state of waterbodies where utilized by population.

TABLE 1. Drainage Areas and Reclaimed Lands in the Main River Basins [4]

River Basin	Catchment Area Belarus km2	Drainage Areas and Reclaimed Lands		Length of Drainage Network in km	
		km2	%	open	closed
Zapadnaya Dvina	33,149	5,142	15.5	20,899	194,418
Neman*	34,610	4,222	12.1	24,830	105,244
Viliya	10,920	1,484	13.5	7,542	43,136
Zapandy Bug	99,994	2,553	25.5	12,082	47,760
Dnieper**	67,545	9,059	13.4	37,489	237,404
Prypyat	50,899	11,287	22.1	52,727	166,156

* Without Viliya

** Without Prypyat

TABLE 2. Resources of ground waters in main river basins of Belarus [4]

River Basin	Fresh Groundwater in km3/year	Withdrawals of Groundwater in 1995 - km3/year
Zapadnaya Dvina	2.97	0.11
Neman	3.51	0.16
Viliya	1.67	0.04
Zapandy Bug	0.66	0.07
Dneiper (without Prypyat)	5.52	0.57
Prypyat	3.57	0.16

TABLE 3. Water Balance Management of Main River Basins for 1995

River Basin	1995 Flow in km3	Reduced Flow From Use km3	Additional Evaporation in Reservoirs and Ponds in km3	Ecological Flows in km3	Balance
Zapadnaya Dvina	15	0.06	0.01	10.5	4.43
Rivers of Zapandy Bug	1.4			0.98	0.42
Neman	6	0.03	0.02	4.62	1.33
Viliy	1.9	0.23	0.03	1.14	0.5
Dneiper	17.6	0.01	0.03	12.7	4.86
Prypat	11.7	0.56	0.07	7.5	3.57

TABLE 4. Sewage Pollution in 1995 [4]

River	Sewage Dumping in Million m3		Pollutants in Tons						
	Natural Surface Waters	Aquifers and Depressions	Oil Prods	SO4	Cl	NH4	NO2	NO3	Cu
Dneiper	1,018	69	250	23,200	36,300	4,055	116	2,499	343
Prypyat	318	27	20	4,100	13,900	371	37	622	120
Berezina	486	13	200	10,800	12,500	2,402	17	1,020	180
Svisloch	308	5	160	500	-	1,982	-	64	118
Sozh	82	15	10	3,700	5,000	104	16	282	7
Neman	188	43	30	12,100	9,600	774	23	789	18
Viliya	35	7	0	700	1,100	114	9	44	6
Zapadnaya Dvina	203	13	40	9,600	2,000	148	30	1,298	4
Rivers of Zapadnaya Bug	52	7	10	1,200	2,500	14	-	-	2
Mychavets	15	4	1	200	500	10	-	-	1

TABLE 5. Long-term Average Flow of Principal Rivers of Belarus Based on B. Fashchevsky's Data

River Basin	River Flows in km3/year	
	Local Points	Entire River
Zapandnaya Dvina - boundary Latvia - Belarus	8.14	13.45
Neman - boundary Lithuania - Belarus (without Viliya)	5.09	6.21
Vilia - boundary Luithuania - Belarus	2.44	2.44
Mukhavets - boundary Poland - Belarus	0.7	0.72
Dneiper - boundary Belarus - Ukraine	11.36	17.62
Prypyat - boundary Belarus - Ukraine	6.1	13.2

3. The State of National Water Resources

As indicated by the changes in the hydrographical network shown in Table 1, the area of reclaimed land reached 12 - 25 percent of the entire catchment area.

The density of open canals associated with small rivers increased 2 - 3 times and if closed drainage networks are included, the density increases 5 - 10 times more [10]. Swamplands decreased sharply, nearly to zero, in some small river basins. The straightening of river channels shortened their length, and, accordingly, increased the slopes of streams by as much as 20 - 60 percent. The straightening of river channels with subsequent increases of stream slopes has caused water levels to drop in the nation's river systems during all phases of the water regime.

For the majority of the small rivers, the low-flow increased 1.5 - 2 times while the annual flows changed by about 5 - 10 percent. The annual flow of the Viliya River changed greatly owing to a yearly volume of withdrawal of about 250 million cubic meters to which must be added losses from reservoir evaporation. We should also mention the degradation of Lake Chervonoye which is situated in the Polessye lowlands. As a result of reclamation and improvement projects, a portion of the lake's catchment area in the Sluch River basin was reduced, and the flood plain was diked and drained by a network of canals[13]. In addition, despite ecological limitations and harm, huge volumes of water are withdrawn from the Zhytkovichsky Canal for the Krasnaya Zorka fish-farm. The withdrawals have caused a 47 cm drop in the water level over the past few years.

These changes in the water regime of the lake broke down the oxygen regime, resulting in an increase in the anoxia phenomenon. Because of the reduced water level of the lake there has been a near complete cessation of flooding in the flood plain. Prior to 1963, the average duration of floods in the plain was 140 days; currently it is down to 12 days.

Consequently, the concentration of magnesium, natrium, potassium, and sulfates has increased two to three times while chlorine concentrates have risen by a factor of one and a half times. Organic pollution (in BOD) increased by twice the normal rate, exceeding the official maximum admissible concentration; even worse, concentrations of copper exceeded what was admissible by a factor of 12.

Available data show (see Tables 2 and 4) that the majority of the principal rivers, lakes, and reservoirs of Belarus are greatly polluted by organic substances such as ammonia, nitrogen, iron, copper, and oil products. Overall, maximum admissible concentrations are exceeded by between 1.5 (at best) to 10 times, particularly as regards copper and oil products.

As can be seen in Table 6, where data for the fallout of Cesium 137 are offered, it is apparent that the fallout drops, depending on the river, by factors of between seven and thirty five from 1987 to 1995[7]. However, it must be stressed that the greatest Cesium 137 activity occurs not in the water but in suspended loads and bottom deposits. The degree of pollution in the water system leads to a second process: exchanges with bottom deposits and the washing out of radionuclides from catchment cover during season of snowmelt and of rain. The largest concentrations of radionuclides in bottom deposits are found in immobile places, e.g., in front of dams and backwaters.

TABLE 6. Annual Cesium Fallouts for the Rivers of Belarus 1987 - 1995[7] (1011 Bq)

Rivers	1987	1988	1989	1990	1991	1992	1993	1994	1995
Besed	8.5	3.7	1.5	0.74	0.56	0.56	0.74	0.814	0.54
Prypyat	29	23	24	7.4	7	7.8	5.4	5.9	3.9
Dnieper	34	23	19	7.8	7.4	4.8	4.8	8.9	4.7
Iput	48	27	4.4	3.7	1.9	1.5	2.1	1.8	1.4
Sozh	100	53	22	11	9.3	4.1	5.3	6	3.1

The greater part of ground water is used for drinking-water supply. The data in Tables 2 and 3 show that Belarus' ground water resources are sufficient for ensuring the needs of the population. However, owing to an uneven distribution of ground water stocks and high levels of pollution, not all settled regions can be ensured good quality drinking water.

Water taken from bore holes situated within the cities of Minsk, Vitebsk, Gomel and Baranovichy, reveal significant increases in minerals, nitrate concentrations, chlorine, sulfates, ammonia and manganese[4]. A rise in ground water pollution in agricultural regions of the country is occurring on stock-breeding farms, at mineral fertilizer storage facilities, in filtration fields, and in polluted areas without a sewage system.

Water utilization by industry and fish-breeding units is declining because of a fall-off of production in both sectors. A different picture emerges regarding the dumping of sewage into natural water sources. Cadastral data for 1995 reveal the sewage was put into natural water systems at specified points at the following rates: 19,000 tons of suspended sediments, 50,000 tons of chlorine, 46,000 tons of sulfates, 16,000 tons of organic of substance, 330 tons of oil products, 310 tons of iron, 35 tons of copper, 42 tons of zinc and large quantities of other pollutants such as phenols and surfactants.

There is, as well, thermal pollution of some Belarus waterbodies that are used as cooling ponds for atomic and thermal power stations, namely the transboundary Lake Drysvyaty which is used as a cooling pond for the Ignalinskaya Atomic Power Station situated in Lithuania, and Lake Lukomlskoye which is used as cooling pond for the Newlukomskaya Thermal Power Station that is sits on the bank of the lake in Belsarus.

These conditions, as revealed by the data make apparent the urgent need for better, more efficient, accurate data collection, archiving, and management if the problems of water planning, policy making, degradation, and distribution are to be solved in Belarus. Solutions will require good distributed database networks and the sharing of data. Experts in Belarus recognize these needs and some have attempted to address them, including the author who has designed a new databases called *Ecological Passport*[12]. Unfortunately, owing to a serious lack of funds for such undertakings, this and other projects have foundered. The *Ecological Passport* database remains empty.

Notes

1. *Yearbook of Hydrological Data for Belarus* (1936-1995) Minsk.
2. *Yearbook of Hydrochemical Data for Belarus* (1965-1995) Minsk.
3. *Data Yearbook of Water Utilization and Sewage Pollution of Rivers in Belarus* (1975-1995) Minsk.
4. *Annual National Water Cadaster for Belarus* (1993-1995) Minsk.
5. *Long-term Data on Water Regimes and Surface Water Resources for Belarus* (1985) Vol. V.
6. *Surface Water Resources of the USSR. The Basic Hydrological Characteristics*, Vol. V., *Belarus and the Upper Dnieper*, (1966, 1974, 1978, 1980) Leningrad.
7. Zhukova, O.M., Matveyenko, I.I., Shuryaeva, H.M., & Chekan, G.S. (1996) "The Peculiarities of Radionuclides Migration In Belarussian Rivers After The Chernobyl Atomic Power Station Accident: The Case of the Iput River, *Proceedings of the International Conference on the Influence of Atomic Power Stations and Other Dangerous Radiation objects on the Hydrological Cycle and Water Resources*, Obninsk.
8. *The Chernobyl Trace in Belarus*, (1996) Minsk.
9. *The Directory of Surface Water Resources: The Surface Water of the USSR*, Vol. V, *Belarus and Upper Dnieper*, (1967) Leningrad.
10. Fashchevsky, B.V., The Impact of Large-Scale Amelioration on the Hydrological Regimes of Belarussian rivers, *Proceedings of an International Conference at the Technical University of Braunschweig, (October 11 - 15, 1993) Germany.*
11. Fashchevsky, B.V., The Evaluation of Ecologically Safe Alterations in the Regimes of Belarussian Rivers, *Proceedings of the International Symposium on Runoff Computations for Water Projects*, (1995) St. Petersburg, Russia.
12. Fashchevsky, B.V., & Shulicka L.G., The Compilation of River Basin 'Environmental Passports' and the Development of Basin Databases, *Proceedings of the International Ecological Congress,* (1996) Voronezh, Russia.

13. Fashchevsky, B.V., Pochodnya G.V., Shatilo A.M., Sheveljuk L.N. An Evaluation of the Influence of Engineering Projects on the Hydrological Regime of Lake Chervonoye, *Proceedings of the All-Union Conference on Anthropogenic Changes of Small Lake Ecosystems*, (1991) St.Petersburg.
14. Fashchevsky, B.V., Fashchevsky T.B., Hydrological and Hydrochemical Changes in the River Basins of the Baltic Sea and the Lakes of Belarus, *Proceedings of the First Conference on BALTEX*, (1995) Visby, Sweden.

WATER RESOURCES PROTECTION AND ECOLOGICAL DATA IN THE REPUBLIC OF KAZAKHSTAN

EDUARD GRANOVSKY, KEMBAEV B.A.,
LYASHENKO L.V., IVASHCHENKO L.A., PANYCHEVA S.B.
Kazak Research Institute of Scientific and Technical Information
221, Bogenbai, Batyr str.
Almaty 480096 KAZAKHSTAN
eduard@gran.ksisti.alma-ata.su

Abstract

This study is based on analyses of *National Reports on the State of the Natural Environment of the Republic of Kazakhstan* in 1994 and 1995 and other official documents. The condition of Kazakhstan's total water resources in the period 1991-95, the impact of pollution stemming from economic development on water resources and the environment, as well as efforts to protect and conserve natural resources together with ecological monitoring by various departments of the Republic are described. The systems employed by various departments and agencies for collecting, processing, and managing information, the drawbacks of the system, the concept of an integrated national data system that has been developed, and the characteristics of databases in the Kazak State Research Institute for Scientific and Technical Information are also described. Finally, information on relevant legislation and supporting documentation on water resources are presented.

Key Words

Aral Sea, data, database, ecology, environment, Kazakhstan, Kazgidromet, Lake Balkash, pollution, systems

1. Water Resources of Kazakhstan

1.1. SURFACE WATERS

The water resources of the Republic of Kazakhstan consist chiefly of river discharges which amount to over 100 cubic kilometers (km^3), 43% of which enters the Republic from the adjacent states of China, Uzbekistan, Kyrgystan, and Russia. The state of resources in 1992-1995 is shown in Table 1.

T. Naff (ed.),
Data Sharing for International Water Resource Management: Eastern Europe, Russia and the CIS, 95–120.

TABLE 1. The Surface Waters Resources of Kazakhstan (km3)

Water Body	Point of Observation	Water Discharge, m^3/s				
		Longterm Average	Average over years of:			
			1992	1993	1994	1995
Black Irtysh	Buran stlm	297.0	292.0	460.0	383.	289.0
Ishym	Akmola City	5.0	12.2	13.1	1.5	---
Ishym	Turgenevka	4.1	---	---	1.8	5.0
Ili	Kapchagai	388.0	322.0	385.0	519.	381.0
Syrdarya	Shardara City	421.0	551.0	671.0	604.	405.0
Ural	Atyrau City	231.0	228.0	378.0	333.	---
Ural	Kushum stlm	307.0	---	---	505.	254.0
Shu	Tashutkul stlm	61.8	42.1	69.0	80.8	48.5
Talas	Pokrovka stlm	20.7	21.1	19.9	34.7	26.0
Assa	Maimak h/s	11.1	10.2	12.3	16.1	10.8

1.2. FRESH WATER ABSTRACTIONS

In 1994, the volume of water in the main rivers of Kazakhstan was above the long-term average, but in 1995 the average flow dropped. In recent years there has been a tendency for lower volumes of abstraction from the rivers nation-wide. For example, total abstractions for 1994 were 1.8 km^3 lower than in 1993. In 1995, practically all indices of water utilization in the Republic of Kazakhstan also significantly decreased.

TABLE 2. Fresh Water Abstractions in Kazakhstan (km^3)

Item	Period				
	1991	1992	1993	1994	1995
Abstracted	36.94	34.12	33.67	31.91	28.81
Used	30.26	27.48	26.92	24.94	22.24
Fresh Water Losses	5.21	5.17	4.99	6.22	5.52
Total Wastes Discharge:	9.00	8.72	8.34	7.49	7.07
- surface water sources	6.88	6.94	6.78	6.04	5.78
- accumulators/depressions	1.68	1.77	1.55	1.64	1.27
- underground water sources	0.44	0.01	0.01	0.01	0.02

Correspondingly, the total water abstractions from natural water sources were reduced by 3.1 km^3 use by 2.7 km^3 fresh water losses fell by 0.7 km^3 waste volumes discharged into surface water bodies fell by 0.26 km^3 and into accumulators or depressions they dropped by 0.37 km^3. The waste volumes diverted into underground horizons made up 0.02 km^3. The bulk of fresh water abstractions are taken from the rivers Syrdarya (31.4%), Irtysh (18.8%), Ili (13.1%), Shu (8.4%) Nura (4.9%) and Lake Balkhash (5.6%).

TABLE 3. Water Abstractions from the Main Rivers of Kazakhstan in 1993 & 1995 (km^3)

Name of the River	Abstracted Total:	including surface waters
Syrdarya	8.33	8.05
Irtysh	5.29	4.82
Ili	3.72	3.37
Shu	2.22	2.22
Nura	1.41	1.26
Balkhash Lake	1.54	1.44
Ural	1.08	-0.98
Torgai	0.32	-0.30
Syrasu	0.34	-0.26

Most of the water withdrawals in 1995 were made in the administrative and territorial divisions of Kyzyl-Orda (4.98 km^3), South Kazakhstan (4.44 km^3), Pavlodar (3.6 9 km^3), Zhambyl (3.66 km^3), and Almaty (3.23 km^3 including the city of Almaty).

1.3. WATER CONSUMPTION

In 1995, total national fresh water demand was 22.24 km^3 of which 20.57 km^3 was supplied from surface waters, 1.67 km^3 from ground water, and an additional 1.19 km^3 came from seawater.

TABLE 4. Total Water Consumption in Kazakhstan 1995

Application	Volume (km^3)	Percent (%)
Industrial needs	4.09	18.4
Domestic needs	1.24	5.6
Irrigation (systematic and basin)	15.80	71.0
Agricultural water supply	0.36	1.6
Pasture flooding	0.33	1.5
Fish farming	0.32	1.4
Miscellaneous needs	0.11	0.5

The largest portion of water is used by the agricultural enterprises of the Ministry of Agriculture (14.37 km^3), energy — Kazakhstanenergo — is the next largest consumer (2.03 km^3), and the Ministry of Power and Coal Industry follows (1.34 km^3). In 1995, 60 percent of industrial water consumption was met by using recycled water. High percentages of recycled water use was noted in The Atomic Power Engineering Corporation (96.4%), The Ministry of Industry and Trade (76.2%), and Ministry of Power Engineering and Coal industry (72.2%). As a whole, the total volume of recycled water use increased between 1994 and 1995 from 0.36 km^3 to 0.06 km^3. Also, water losses during transport decreased considerably for the same period, from 6.22 km^3 in 1994 to 5.52 km^3 in 1995.

1.4. WASTE WATER AND POLLUTION

In 1995, the total volume of wastewater was 5.78 km^3. Of this amount, wastewater of standard purity (i.e., without artificial purification) was 5.29 km^3, that treated to standard purity was 0.26 km^3, and polluted water (without purification or insufficiently treated) was 0.23 km^3. All these waters were diverted to surface water sources. The volume of polluted water has declined recently.

TABLE 5. Polluted Water Discharges

Years	1991	1992	1993	1994	1995
Polluted wastes (km^3)	0.34	0.31	0.29	0.24	0.23

There are over 5400 water-consuming enterprises in the Republic of Kazakhstan. The principal dumpers of wastewater into natural watersources are Kazakhstanenergo (1.8 km^3), Goskomvodresursy (1.1 km), and the atomic energy and other industries (1.1 km^3). The regions where the largest volumes of industrial discharges of polluted water occur are the territories of East Kazakhstan (55.0%), Pavlodar (22.8%), Karaganda (11.2%) and Semipalatinsk (10.1%).

The volume of collector-drainage waters in the "standard purity" category that were diverted from irrigated lands in 1995 decreased by 0.48 km^3 as compared to 1.5 km^3 in 1994. The main volumes of this category of water are diverted from the territories of three regions: South Kazakhstan (0.65 km^3), Kyzyl-Orda (0.47 km^3) and Almaty (0.28 km^3).

Despite the considerable decrease of polluted waste discharge, the quality of watercourses nationwide remains unsatisfactory. In particular, the Irtysh River basin was damaged ecologically because of 0.202 km^3 of polluted wastes which constituted 88% of the River's total discharge.

Between 1994 and 1995, the levels of contaminants detected in the nation's rivers changed considerably. In 1995, 10 times more suspended matter and 2.2 times more compounds of iron were discharged into watercourses, while the discharge of compounds of copper, zinc, nickel, and mercury decreased considerably.

TABLE 6. Levels of Contaminants in Surface Waters

Name of Substances	Measurement Unit	1994	1995
Suspended matter	1000 tons	250.66	25.19
Ammonia nitrogen	1000 tons	2.91	2.71
Nitrogen of nitrate	1000 tons	1.91	0.71
Organic compounds specified by BOD6	1000 tons	6.96	5.95
Oil products	1000 tons	0.23	0.15
Phenols	tons	0.57	0.29
Surface active substances	tons	84.01	93.18
Phosphor	tons	51.39	55.86
Compounds of iron	tons	204.63	91.28
Compounds of copper	tons	7.12	8.65
Compounds of zinc	tons	24.89	40.22
Compounds of nickel	tons	0.05	1.49
Compounds of mercury	kg	35.70	64.40

1.5. QUALITY

According to the comprehensive estimates of Kazgidromet the Ural and Irtysh Rivers in whose water pollution indices (WPI) stood, respectively, at 7.18 and 6.56 units (Table 7) are considered to be the most polluted rivers in the nation as measured by a group of specific chemicals. The level of water quality in all of the remaining rivers, including the smallest, have remained at the level of 1994. WPI is determined by the level of particular contaminants specific to each basin.

TABLE 7. Quality of the Water Sources of Kazakhstan

Basin Name	Specific Substances	WPI By Units			
		1992	1993	1994	1995
Ural	Oxygen, BOD_5, oil products, phenols, boron, copper	2.8	2.5	2.55	7.18
Syrdarya	BOD_5, nitrogen of nitrite, copper, oil products, phenols	1.4	0.82	0.75	1.60
Nura	Nitrite and ammonia, nitrogen, oil products	2.7	2.2	2.9	1.60
Sarysu	Oil products, copper, phenols, nitrite and ammonia nitrogen	2.5	2.94	3.83	3.60
Ili	BOD_5, nitrite nitrogen, oil products, phenols, fluorine, copper	2.0	1.33	1.71	1.32
Irtysh	BODs, oxygen, nitrite and ammonia nitrogen, copper, zinc, oil products, phenols, xanthates	10.4	7.4	8.11	6.56
Ishim	Iron, sulfates, oil products	1.2	1.2	1.58	1.24

1.6. DISTRIBUTION, USE, AND POLLUTION

Irtysh River Basin. The Irtysh River is the principal water artery of Kazakhstan. It supplies water to the eastern region of the Republic where over 900 industrial water consumers, encompassing such industries as non-ferrous metallurgy, chemical, machinery, oil refining, food processing, and others, are situated. Withdrawals from the basin's surface sources in 1995 made up 4.82 km^3 (20% of the country's total abstractions). Three hundred thirty three enterprises in the Pavlodar region of the basin withdraw 3.64 km^3 of fresh water. In the East Kazakhstan and the Semipalatinsk regions water abstractions were, respectively, 0.51 km^3 and 0.36 km^3.

Total waste discharge into the river basin amounted to 2.1 km^3, including 0.20 km^3 of polluted wastes. Most of the polluted wastes (0.13 km^3) flowed into water intakes from the territory of East Kazakhstan region where in 1995 ninety-four enterprises with independent wastewater outlets functioned. Ten industries in Pavlodar and 12 in Semipalatinsk discharged 1.79 km^3 of waste into the Irtysh River basin of the river Irtysh, of which 0.07 km^3 were polluted.

Water quality over the entire length of the Irtysh River basin, together with its tributaries, has been assessed as "very polluted." In 1995, 16 cases of extremely high pollution level were registered, particularly in the Glubochanka and Krasnoyarka Rivers owing to very high levels of copper and zinc stemming from the failure of pumps, the pumping stations, the drainage system, and the stoppage of the lime supply.

Local and city purification facilities do not provide a satisfactory level of wastewater purification. Consequently, in 1995 these units discharged 43.2 thou t of suspended matter, 24.3 t of zinc compounds, 8.06 t of copper compounds, 0.31 t of chromium compounds and other substances into the water body of the Irtysh River basin.

The Syrdarya River Basin. The Syrdarya flows through South Kazakhstan and the Kyzyl-Orda regions which are Kazakhstan's main rice and cotton growing areas. In 1995, 8.03 km^3 of fresh water were abstracted from the surface sources of the Syrdarya basin — 3.12 km^3 from South Kazakhstan and 4.91 km^3 from the Kyzyl-Orda region. The total capacity of sewage works in the two regions 0.21 km^3, whereas 0.76 km^3 of wastes was discharged into the waters of the Syrdarya basin.

Collector-drainage wastes are classified as "of standard purity" and are directed into water intakes without any purification. In 1995, in the South Kazakhstan region, 0.64 km^3 and in Kyzyl-Orda region, 0.12 km^3 of collector wastewater was discharge into the basin. The Syrdaray River is classified as "moderately polluted." In 1994, however, the collector-drainage waters, contaminated with considerable amounts of pesticides and mineral fertilizers from rice-growing and other kinds of farming, and ranging in content from 10 to 50 mpc were introduced into the estuary of the Syrdarya. The quality of Syrdarya water deteriorated in 1995, with the WPI rising to 1.6 as compared to 0.75 in 1994.

The Ili River Basin. Ili River and its tributaries flow through the territory of Almaty and the Taldy-Korgan regions where the waters of the river are used mainly for agriculture. In 1995, fresh water withdrawals from surface sources were 3.37 km^3, of which 2.34 km^3 were consumed, and 0.33 km^3 of waste water, including 0.32 km^3 of water of standard purity without purification were discharged into the basin.

The Ili-Balkhash Basin. All of the water systems of the Ili-Balkhash basin flow into Lake Balkhash. In 1995, withdrawals from the lake — 1.44 km^3 — decreased by 0.25 km^3 Main users of the lake's waters are the non-ferrous metallurgy industries and agriculture. Together, they consume 1.20 km^3 of water. In 1995, wastes in the amount of 0.19 km^3, including 0.0017 km^3 categorized as polluted, were discharged into Lake Balkhash.

The Nura River Basin (Central Kazakhstan). This is a low water basin owing to the highly regulated flow of the river which receives considerable waste discharges from industry. In 1995, fresh water abstraction from surface sources totaled 1.26 km^3, of which 1.53 km^3 were consumed and, 1.29 km^3 were discharged as waste. Annually, 1.03 km^3 of unpurified wastewater is abstracted from the water sources of the Nura basin. The quality of Nura water was assessed in 1995 as "moderately polluted." In 1994, the rivers Nura and Sherubai-Nura, receiving the wastes of metallurgical and petrochemical industries, passed into the category of "secondary pollution" with mercury. Pollution level of elastic deposits reached 1.2 m, and the weight of the accumulated mercury reached several tens of tons.

The Ural River Basin. In 1995, 0.93 km^3 of water was abstracted from surface water sources. The majority (0.52 km^3) was used for irrigation and pasture flooding, and 0.02 km^3 of wastes including wastes "of standard quality" (0.0041 km^3) and the rest, "treated to standard," were discharged into the surface waters of the basin. The pollution index increased from 2.55 in 1994 to 7.18 in 1995, as a result of which the water quality of the river was assessed as "very polluted."

Within the territory of Kazakhstan, there is no discharge of polluted wastewater into the Ural River. The chemical composition of water in the upper stream is constituted as a result of contaminants coming from the territory of Russia. Old slime settlers and chrome compounds of the Aktyubinsk chemical plant pollute the river with compounds of fluoride, boron and chromium and have an impact upon the quality of surface waters of the river through the underground sources. An increase of these substances has been observed during the autumn and winter periods.

1.7. GROUNDWATERS

In Kazakhstan, 609 deposits of groundwaters (GW) are exploited for domestic, industrial, service, and irrigation uses. Underground waters are used at the rate of 15.76 km^3 per year. Kazakhstan's underground waters, like its surface sources, vary widely in distribution. More than 63% of the nation's groundwaters are concentrated in the southeastern regions of the country. The Atyrau, Mangistau, Torgai, Akmola, Kokshetau, North Kazakhstan and West Kazakhstan regions are relatively poor in groundwaters. In fact, for all practical purposes, Atyrau, Aktau, Pavlodar, Petropavlovsk, Semipalatinsk may be considered to have no groundwaters.

The largest abstractions of groundwater are made in the Almaty, South Kazakhstan and East Kazakhstan regions. In 1995, groundwater withdrawals were 2.16 km^3, of which 57.8% was used for domestic water supply, 18.0% for industrial-technical purposes, 7.3% for irrigation, and 6.5% for pasture flooding.

As a result of intense groundwater abstraction and mine drainage there was not only a considerable regional decrease in the level of groundwater, but there was significant earth surface settling as well. Great depression cones, in some cases reaching more than 1000 km^2, and the level of decrease (in the center of the depression) reaching 100-150-500m in South-Kazakhstan, Zhezkazgan, Karaganda, Kostanai, Kokshetau, and the East Kazakhstan regions were observed.

As in previous years the groundwater resources were subject to great anthropogenic pollution. As of 1995, 456 groundwater pollution sources were revealed, 103 of them are systematically observed. The content of pollutants in groundwaters is given in Table 8.

The primary substances polluting groundwaters are sulfates, compounds of nitrogen, phenols, and heavy metals. In the areas where metallurgical industrial enterprises are located, cadmium, selenium, and mercury are present in groundwaters.

Due to the fact that the list of the substances to be controlled is not regulated, the characteristics of groundwaters pollution can only be roughly estimated. Much higher levels of groundwater protection from technogenic pollutants, as compared to surface waters, determine the advantages of their use for domestic water supply.

TABLE 8. Pollutants in the Ground Waters of Kazakhstan

Name of Regions and Enterprises	Pollutants	MAC Exceeding Factor	Source Sizes km2
KARAGANDA REGION JSC Ispat-Karmet, PA Karbid Interfluve Sherubai-Nura Sokyr (mine waste waters)	phenals mercury nitrates mineralization	20-79 22-134 5.0 15.0	
ZHAMBYL REGION Slime accumulator and settlers ZhPA CHIMPROM Superphosphate plant	nitrates fluorine nitrates	1.7 8.2 3.0	12
PAVLOSAR REGION Pavlosar section of the north industrial center (chemical, ruberoid, tractor plants) Slime accumulator of Ermakov GRES Ferroalloys plant	mercury fluorine ammonium fluorine arsenic	4.0-7.4 0.1-1.9 1.5 0.8-5.0 0.1-0.4	200
AKTYUBINSK REGION Aktyubinsk industrial zone, heat plant northeastern direction downstream, thinning out In Ilek river	chromium	7-294	11
EAST KAZAKSTAN REGION Ust-Kamenorgorsk, industrial zone	mineralization nitrates nitrates sulfates chlorides copper lead zinc manganese fluorine cadmium selenium	8.9 9.0 3.3 11.7 1.6 33.0 11.0 55.4 23.0 4.2 12.0 48.0	27.5
ALMATY REGION Almaty GRES	oil products zinc aluminum	8.1 3.6 10.0	

1.8 FRESH SEA WATERS

Three unique intercontinental water bodies: the Aral Sea, the Caspian Sea and the Lake Balkhash are located on the territory of the Republic of Kazakhstan. A retrospective hydrogeology of these seas is given in Table 9.

TABLE 9. The Hydrogeological Characteristics of Kazakhstan's seas.

Water Body Name	Observation Period	Level abs.m	Water table area, thou. km^3	Water volume in water body, km^3	Average volume of tributaries km^3
Aral Sea including "Large Sea"	1953-1961	53.5	61.1	1080.0	
	1981-1986	50.8	60.0	922.0	
	1987-1991	37.4	31.2	246.0	55
	1992	36.9	30.9	232.5	46
	1994	36.6	29.6	223.8	
	1995	36.8	30.2	232.0	
"Small Sea"	1987-1991	38.0	2.5	16.7	
	1992	38.0	2.9	22.0	
	1993	38.3	2.5	22.0	
	1994	40.0	2.8	22.0	
	1995	40.0	2.8	22.0	
"Caspian Sea"	1991	-27.2	380	78557	385
	1992	-26.9	392.7	78652	287
	1993	-26.9	393.2	78668	340
	1994	-26.7	395.7	78747	396
	1995	-26.6	397.2	78780	*
Lake Balkhash	1991	341.3	16.9	92.56	13.3
	1992	341.1	16.6	89.62	13.2
	1993	341.1	16.5	88.78	18.2
	1994	341.2	16.9	92.04	20.3
	1995	341.3	17.1	93.4	14.4

TABLE 9 (cont.)
* no data

The Aral Sea is a body of water with features peculiar both to a sea and a lake. During the past forty years the Aral Sea streamflow has been reduced approximately five times, the area of the Aral's water surface has decreased more than twice, the level of water in the sea has decreased 1.5 times, and the influx of river flow has ceased entirely since 1987. In 1993-1994, the ecological state of the Aral Sea continued to deteriorate in practically all parameters, and in 1995 it did not improve.

The Caspian Sea is the largest body of water body on the planet without an outflow. The total length of the Caspian Sea coastline is 7,000 km, including 2,300 km within Kazakhstan. The cyclic change in the level of the Caspian Sea is determined mainly by climate. Starting in 1978, there has been a gradual rise in the level of the Caspian which continues today with at an average of 0.14 m/yr. In the zones where flooding occurs due to the sea's rise, populated areas, agricultural lands, and oil deposits continue to be affected. The ecological condition of the sea also continues to deteriorate; it is categorized as moderately polluted by Kazak standards.

Lake Balkhash, whose water level in late 1995 was 341.3 abs.m, has not changed in this regard. The water level in Lake Balkhash has not changed of late — in 1995 it made up 341.3 abs.m. In fact, despite a reduction in the incoming river flows in 1995 as compared to 1994, the total volume of water in the lake actually increased slightly to 93.4 km^3. However, the quality of water in Lake Balkhash remains, as it has previously, unsatisfactory. The content of heavy metals, oil products, and phenols exceed the standard requirements by 40.3%, 38.2%, and 24.6% respectively. The most polluted section of the Lake Balkhash is the region of the Bay Bertys, from which originates the industrial wastes of the copper smelting, the electrolytic, and non-ferrous metals plants of Pa Balkhashmed. The average copper concentration in this part of the lake reached 26 MAC in 1994.

1.9 DRINKING WATER SUPPLIES

The situation regarding the delivery of safe drinking water to the population has been and remains tense. In the entire nation, the average supply to cities and other populated areas is below the permissible standard by a factor of approximately three and six to eight times in some of its regions. In 1995, the severity of the problem hardly decreased. Table 10 cites the indices of the specific water consumption of the population in all regions of the country and the technical level of water supply sources.

More than 80% of the urban populations of southern Kazakhstan, Aktyubinsk, Atyrau, Kostanai, Zhezkazgan, and eastern Kazakhstan, together with the capital city of Almaty were better supplied with domestic water in 1995. The most poorly supplied populations were in the urban settlements of the Kyzyl-Orda, northern Kazakhstan, Torgai, and Taldykorgan regions. More than 40% of the rural population of the of Kazakhstan use water for domestic purposes from wells, springs and other non-piped water sources.

TABLE 10. Kazak Domestic Water Supply
(Water consumption: urban 550 liters/capita; rural 125 liters/capita)

Name of Region or City	Specific daily average water consumption (l/cap)		Water supply sources - %			
	City	Village	Tap water	Non-Piped sources	Open basin	Transported water
REPUBLIC OF KAZAKHSTAN	25-500	15-320	81.3	11.9	1.6	3.0
Almaty regions	251.0	–	100	–	–	--
Aktyubinsk	250	40-120	81.4	16.5	0.3	1.8
Akmolinsk	125-160	35-60	87.9	4.3	–	2.5
Almaty	80-140	35-50	79.5	21.7	3.7	2.3
Atyrau	250	50-320	73.5	6.9	12.6	7.0
E. Kazakhstan	150-300	30-125	85.3	12.1	1.3	0.2
Zhambyl	105-290	27-164	78.7	18.0	–	3.3
Zhezkazgan	40-450	25-75	84.0	10.1	0.1	2.9
W. Kazakhstan	130-150	15-80	56.9	3.8	1.5	7.6
Karaganda	120-200	70-125	94.5	4.4	0.1	0.5
Kakshetau	150	150	63.0	27.0	–	10.3
Kostanai	250	60	80.3	14.7	–	2.6
Kyzal-Orda	50-140	20-40	87.7	9.1	0.4	2.3
Mangistau	120+300	70+250	80.0	0.4	–	19.6
Pavlodar	220-240	50	78.7	16.3	0.3	4.7
N. Kazakhstan	25-250	25-30	92.2	5.1	1.6	1.1
Semipalatinsk	40-200	30-120	76.0	25.0	–	0.8
Taldykorgan	60	60	87.6	2.8	9.4	5.6
S. Kazakhstan	60-500	35-80	83.0	9.9	3.6	3.2

The quality of the greater part of domestic water supplies, both surface and ground water, does not correspond to GOST standards. In 1995, the quality of water consumed by the population over the entire country deteriorated in terms of bacterial pollution but improved as regards chemical indices and sanitary conditions (Table 11).

TABLE 11. Sanitary Conditions of Kazak Water Sources

Indices	**Quantity of water samplings from the tested not meeting the hygienic norms (%)**					
	1990	**1991**	**1992**	**1993**	**1994**	**1995**
Sanitary-chemical	13.4	20.0	14.2	16.7	14.4	11.7
Microbiological	19.2	14.2	9.4	9.2	8.6	9.7
Including those containing lactose positive colibacillus			57.6	57.6	66.0	64.8
Isolating pathogens of infectious diseases				29.0	3.4	3.6

The most unsatisfactory quality of the tap water is recorded (in %) in the regions of Kokshetau - 15.6% (in 1994 - 13.8%), Kyzyl-Orda -11.8% (10.6% in 1994), Torgai - 9.9% (13.2% in 1994) and Almaty - 8.2% (4.1% in 1994). In 1995, the sanitary state of unpiped domestic water sources in terms of bacterial count deteriorated over the Republic as a whole (Table 12). The portion of water samples not meeting sanitary norms increased by 2.4% in 1995 as compared to 1994. The sanitary state of unpiped sources of the Torgai, Kostanai, Kokshetau, Akmola, and Zhambyl regions significantly deteriorated.

TABLE 12. Condition of Unpiped Domestic Water Supply by Bacterial Count

Region	**Share of water samples not meeting sanitary norms (%)**	
	1994	**1995**
REPUBLIC OF KAZAKHSTAN	9.1	11.5
Torgai	12.4	32.9
Kyzyl-Orda	27.8	32.9
Kostanai	15.7	21.0
Kokshetau	15.5	17.1
Shymkent	18.1	15.4
Akmola	2.9	9.6
Zhambyl	6.9	13.1

The unsatisfactory state of domestic water supply systems was responsible for the spread of infectious diseases transmitted by water. The chief causes of the poor quality of the domestic water supply systems are:

- discharge of polluted, insufficiently purified wastes into water sources
- low technical quality of domestic water supply systems
- low levels of sanitary controls and absence of disinfectant facilities and other preventive measures

- unlawful "cutting in" into water supply system which causes secondary water pollution
- absence of chlorine-containing and coagulating agents for water treatment.

2. The Impact of Economic Activities on the Condition of Water Resources

2.1. GENERAL INFORMATION

Gross agricultural output for the entire country decreased by 21% in 1995. Consequently, the load on the environment substantially decreased. Thus, as compared to 1994:

- effluents from stationary sources (395,000 tons) decreased by 5% motor transport effluents (997,000 tons) decreased by 17%
- fresh water diversion from natural sources (28,807 mcm) decreased by 9.8%
- discharge of polluted wastes (230 mcm) into natural water sources decreased by 2.6%.

In 1995, the total volume of polluted wastes discharged by the three main economic sectors into surface water sources of the Republic of Kazakhstan were:

- industry (63.4%)
- agriculture (32.5%)
- housing facilities and public utilities (4.1%).

At the beginning of 1996 more than 20 billion tons of industrial wastes (toxic ones inclusive) which continue to be stored in diverse accumulators without observing any ecological norms and requirements are accumulated in the republic. As a result, ground and surface water resources in many regions of the Republic are subject to intense pollution. On the whole, in 1995 there was no decrease of the technogenic load upon the environment as compared to 1994.

2.2. THE POWER INDUSTRY

Of the total volume of effluents produced by industrial sources in 1995 —3, 097, 000 tons — 1, 099,000 came from the power industry, of which 42.7% were solid wastes. Gaseous wastes were made up of about 64% sulfur dioxide, 16% of carbon oxide, and 18% nitrogen oxide. Waste discharges into surface waters increased by 1,689 mcm in 1994 and by 2,920 mcm in 1995. Suspended matter, oil products, chlorides, sulfates, and the salts of heavy metals are discharged into natural features together with the wastes of power plants.

Ash and slag dumps of power plants that produce heat and electricity continue to be important sources of soil and ground water pollution. About 15 million tons of ash and a combination of ash and slag wastes per year are formed in the industry. By 1995 more than 300 million tons of ash and slag wastes had accumulated in the Republic.

2.3. FERROUS METALLURGY

In 1995 ferrous metallurgy industries discharged into the atmosphere 3,490,000 tons of harmful substances, an amount 0.6% less than in 1994. Plants of the Karaganda, Kostanai, Pavlodar, and Aktyubinsk regions discharge the greatest volumes. In 1995 455 mcm of ferrous industrial wastewaters were discharged into the countries surface water sources.

2.4. NON-FERROUS METALLURGY

Non-ferrous metallurgy plants are concentrated chiefly in east Kazakhstan and in the Zhezkazgan regions. The specific pollutants of non-ferrous metallurgy are: copper, zinc, lead, arsenic, mercury, sulfuric acid, hydrogen fluoride and others. In 1995, these industries pumped some 71,621 mcm of waste into the surface waters of these regions, about 1.9 times more than in 1994. The main pollutants of these water resources are copper and zinc. The non-ferrous metallurgy enterprises accumulated a great amount of toxic industrial waste, the main part of which make up slags and flotation tailings. In the east Kazakhstan region alone there had accumulated by 1995 some 18,519,000 tons of toxic wastes in solid, liquid and pasty form.

2.5. THE CHEMICAL INDUSTRY

In recent years the volume of effluents, discharges, and wastes produced by the chemical industry decreased considerably —a drop of 13% between 1994 and 1995 — due, in greater part, to production recession. In 1995, 58.9 thou t of harmful substances were discharged into atmosphere, that is 13% less than the level of 1994. Despite the considerable decrease of waste discharges by the chemical industry, they continue to exert a negative impact upon the natural environment. The mineral fertilizers produced by chemical plants contain such highly toxic substances as phosphoric anhydride and hydrogen phosphide. In 1995, the chemical industry accumulated about 2,350,000 tons of toxic wastes in solid, liquid and pasty state. Some plants discharged 119 kg of mercury and 4380 kg of hexavalent chromium. Chemical industry wastes are discharged mainly into accumulators.

2.6. THE OIL AND GAS INDUSTRIES

In 1995 the total volume of wastes in the oil and gas industries amounted to 177 thousand tons. In 1995, 2.72 mcm of unpurified wastes from these industries were dumped into natural surface waters, 23.51 mcm into accumulators and depressions, and 0.24 mcm into underground sources.

The efforts of JSCs, and the associations of MUNAIGAS to decrease the negative impact of economic development on the natural environment resulted in a decrease of the discharge of harmful substances into the atmosphere by 68 thousand tons, a 37% drop over 1994 figures, a reduction of fresh water consumption by the oil and gas industries of 4.6 mcm, a 4% drop, and a decrease in the discharge of polluted wastes into water sources by 1.0 mcm, a reduction of another 4%.

2.7. THE CONSTRUCTION INDUSTRY

In 1995 the industry used 15.91 mcm of fresh water, 28.8% of it utilized for industrial needs. Of this total, 0.04 mcm of polluted wastes containing suspended matter, oil products, ammonium nitrogen, nitrates, phosphorus, magnesium, iron and others were discharged into water sources. Annually, over 3.1 thousand tons of toxic wastes containing asbestos, a small part of which is utilized, are accumulated by the construction industry, 84% of which are transported to organized storage sites.

2.8. HOUSING FACILITIES AND PUBLIC UTILITIES

This sector exerts a negative impact on the natural environment because of its use of surface and ground waters for domestic, potable and industrial purposes, and because of its discharge of unpurified or partially purified domestic and industrial water supplies, as well as surface run-offs and the atmospheric emission of boiler houses, and the disposal of domestic and industrial wastes at dumping sites.

The country's municipal facilities and services are among the largest users of fresh water, (about 1000 mcm/yr), 35% of which is replenished at the expense of surface waters and 65% at the expense of ground waters. The quality of ground waters used for water supply for the most part meets the established sanitary requirements, but their pollution with oil products, heavy metals, and pesticides is steadily increasing. Each year, 287mcm of wastewater — 45% of which is of standard purity and 54% not fully purified — are discharged into surface waters through communal sewerage systems.

2.9. AGRICULTURE

Due to a recession in agriculture in, 1995, the negative impact of mineral and organic fertilizers and pesticides upon the environment decreased. Agriculture consumes 80%, or 2,096.1 mcm, of the total volume of water abstracted from the national supply. About 25% of this amount is lost in delivery. 1,556,000 mcm of fresh water is used for irrigation, of which 1,205,000 mcm are discharged into watercourses and accumulators, 418 mcm into depressions and reliefs.

2.10. An Overview of the Impact of Economic Development on the Environment

TABLE 13. Discharge of Wastes into Surface Waters (mcm)

ECONOMIC SECTORS	1994	1995
Republic of Kazakhstan	6035.91	5780.83
Industry as a whole including:	4305.33	4284.36
- power industry	1231.47	2920.22
- non-ferrous metallurgy	36.94	71.21
- oil and gas	1.94	2.72
- construction	0.50	0.04
- housing facilities and public utilities	274.71	287.10
- agriculture	1466.17	1204.93
- other branches	4.72	4.44

TABLE 14. Discharge of Atmospheric Pollutants (1000 tons)

ECONOMIC SECTORS	1994	1995
Republic of Kazakhstan	3261.0	3097.4
Industry as a whole including:	2529.0	2216.7
- power industry	1242.0	1098.5
- ferrous metallurgy	370.2	348.6
- non-ferrous metallurgy	576.0	502.9
- chemical industry	67.5	58.9
- oil and gas	172.0	177.2
- construction	59.0	30.6

2.11. FUNDING ENVIRONMENTAL PROTECTION

In 1995, 5.9 billion tenge, in comparable prices 42% less than in 1994 (table 15), was expended by the Republic for environmental protection and optimal uses of natural resources by all forms of industry and organizations.

TABLE 15. Sources of Funding and Capital Investment for Environmental Protection (millions of tenge)

Sources of financing	Capital Investments	
	1994	1995
Total budget	674.0	275.2
- Central gov't		229.7
- Local gov'ts		45.5
Funding by industrial enterprises	6741.0	4947.2
Combined investments	7415.0	5222.4

The share of capital investments for environmental protection decreased from 9.1% in 1994 to 5.3% in 1995, while the share of the industrial sector increased from 90.9% to 94.7% in the same period. In terms of absolute monetary values, the decrease was 2.5 and 1.4 times, respectively.

TABLE 16. Distribution of Investments for Environmental Protection in 1995 (millions of tenge)

Types of Expenditure	Capital Investments	
	Tenge	%
Protective infrastructure	3399.5	65.1
Water resources conservation	913.9	17.5
Air protection	647.6	12.4
Forest, mineral resources, fish reserves	261.4	5.0
TOTAL	5222.4	100

The great bulk of the investments for environmental protection, amounting to about 53% of the national total, were expended, as in previous years, in the regions of Karaganda, Pavlodar and Atyrau regions. On the other hand, the regions of Very Kyzyl-Orda, Zhambyl, Taldykorgan, Torgai, Semi palatinsk, Kokshetau, Kostanai and the city of Almaty received very insignificant investments.

TABLE 17. Regional Distribution of Investments for Environmental Protection (millions of tenge)

Region	Capital Investment	
	tenge	%
Republic of Kazakhstan	5222.4	100.0
Karaganda	1084.2	20.8
Pavlodar	862.4	16.5
Atyrau	798.2	15.3
Zhezkazgan	475.2	9.1
Mangistau	427.6	8.2
East Kazakhstan	407.9	7.8
Aktyubinsk	376.8	7.2
Almaty	304.3	5.8
South Kazakhstan	132.8	2.5
West Kazakhstan	70.4	1.3
North Kazakhstan	66.4	1.3
Akmola	52.4	1.0
Kostanai	44.0	0.8
Kokshetau	30.0	0.6
Semi palatinsk	28.9	0.6
City of Almaty	28.0	0.5
Torgai	26.2	0.5
Taldykorgan	3.7	0.1
Zhambyl	2.0	0.1
Kyzyl-Orda	1.0	0.02

TABLE 18. Capacity of Environmental Protection Projects (1000 m^3/day)

Completed Projects	Capacity	
	1994	1995
Waste treatment stations	28.1	60.9
Water recycling supply systems	758.4	1.7

In addition, facilities for the removal and processing of harmful substances from waste gases had a capacity (in thousands of tons per day) of 15.0 in 1995 (no figures available for 1994).

In 1995 the level of expenditure on the preservation of the environment increased significantly. Just short of 11 billion tenge were spent for water conservation (6.80 billion), air protection (3.79 billion), and the reclamation of land (0.28 billion). Regionally, the largest share of expenditures were made in East Kazakhstan (2.13

billion) and Pavlodar (2.03 billion), with the least amount spent in Almaty (0.086 billion), Taldykorgan (0.068 billion), Torgai (0.033 billion), and Kyzyl-Orda (0.032 billion).

3. Ecological Monitoring and Information

Ecological monitoring in Kazakhstan is carried out by various government ministries, committees, and administrations.

TABLE 19. Ecological Monitoring Activities of State Organizations

STATE BODY	FIELD OF RESPONSIBILITY
Ministry of Ecology and Biological Resources	Pollution sources, bioresources, reserves radio-ecological conditions
Hydrometeorological Administration	Atmosphere, surface water, soils, radiation, background monitoring
Ministry of Geology and Mineral Resources	Ground water resources
Ministry of Public Health	Health & contagious infections, drinking water, food, sanitation
Goskomzem	Earth and soil
Atomic Energy Agency, Ministries of Public Health & Ecology & Biological Resources	Radiation (incl. industrial noise and electromagnetic levels
Emergency Committee	Emergencies, natural and anthropogenic disasters
Water Resources Committee	Surface waters, floods
Ministry of Science /Academy of Sciences	Seismic & scientific monitoring

Source: ME& BR Report and the firm of SAIS, Almaty

Diverse ministries and departments implement government regulation of environmental conditions. The hydrochemical content of surface waters is controlled by subdivisions of Kazgidromet. Government laboratories established for sanitation testing and epidemiological inspections monitor the quality of piped and unpiped drinking. These laboratories are also responsible for testing the quality of surface waters and reservoirs at points of use. The Monitoring Center the Ministry of Geology and Preservation of Mineral Resources is responsible for ground water quality and harmful geological processes. The agencies of the State Sanitary Inspection Unit track residual nitrates in food plants and pesticides in foodstuffs.

The systematic observation of surface water quality in Kazakhstan was begun toward the end of the 1930s and beginning of the 1940s. Presently, control of the chemical and physical qualities of surface waters is implemented through 295 control stations in 83 rivers and reservoirs. Hydrobiological contamination is monitored in 32 watercourses at 153 control stations. Kazgidromet laboratories measure over 70 water quality indices.

Observations of seawater pollution is implemented in the Caspian and Aral seas in accordance with up to 30 quality indices. The departments of laboratory analysis and control (DLAC) of the various regional and city administrations are responsible for analyzing the condition of water resources, discharges in to reservoirs, the operation of sewage from various industrial enterprises, atmospheric emissions, soil pollution, and information, thus having "supra-departmental" control over these activities.
In 1996, the DLAC subdivisions controlled 1,987 enterprises for water sources pollution and made 3,753 inspections during which 11,743 samplings and 141,344 determinations were made.

3.1. MONITORING AND DATA COLLECTION

A continuous process of generating data results from the monitoring of the state of water resources. At the outset of the 1980s, the processing of hydrological observations was carried out with the help of an automated information system (AIS) known as "Hydrochemistry," using EC computers. Pollution data from the monitoring activities has been stored in departmental data collections and published. Beginning in 1996, personal computers have been used in Kazakhstan in the processing and analysis of data using AIS AQUA, and the programming language FOXPRO and a database for the processing and management of data has been created. The database consists of:

- list of monitoring stations
- measurments of hydrochemical pollution
- hydrochemical composition
- complex estimates
- reviews and references
- background value
- water content characteristics
- quantity of samples.

The *State Water Cadastre. Annual Data on Inland Surface Water Quality* is published from the data in this database. Together the water resources database, two others documentary and bibliographic databases on the Caspian Sea and the Aral Sea have been created. These databases contain information on research and development reports, periodical articles, manuscripts on hydrometeorology and on the pollution of water. The documentation in the Caspian Sea database goes back to 1964 and consists of 607 documents, and the Aral Sea documents date back to 1979 and number 514 documents. The Kazak State Research Institute of Scientific and Technical Information (KazgosINTI) houses the National Information Center contains published and unpublished papers of Kazak scientists and other specialists, scientific abstracts and bibliographies on water and pollution that date back to 1991. There are also a number of data collections that contain information on these problems, e.g.:

- Listings of databases in Kazakhstan since 1993 (over 200 documents)

- Scientific and technical programs carried out in Kazakhstan since 1993 (about 400 documents)
- Research projects carried out in Kazakhstan since 1995 (over 3,700 abstracts and bibliographical items)
- Dissertations defended in Kazakhstan since 1994 (over 1700 documents)
- Research projects deposited in KazgosINTI since 1994 (over 1,100 items)
- Research and development projects and production experiences of Kazakhstan since 1994 (about 10,000 documents)
- Scientific and technical literature processed by the automated system since 1992 (over 30,000 documents)
- National Register of Patents for inventions since 1993 (Patent Office of the Republic of Kazakhstan — over 4,100 documents).

KazgosINTI is the headquarters for the Development of the national system of scientific and technical information (1993-1998). One of its targets is the establishment of a distributed database system which would include ecological databases. The Ministry of Ecology and Biological Resources in cooperation with other concerned ministries issues a quarterly bulletin entitled "the Ecological Information Bulletin" which serves the interests of various government managers and informs the public. Special information (reference information, bulletins and so forth) on state of the natural environment and its pollution are given to users on a subscription payment basis.

3.2. LAWS GOVERNING THE ENVIRONMENT AND WATER RESOURCES

At present, the draft law on hydrometerological activity, whose aim is to meet the needs of state power agencies, the Emergency Committee, and legally authorized and non-official persons in regard to information and data on the condition of the environment is the responsibility of the Agency for Hydrometeorology and the Monitoring of the Environment. In addition, the draft law is intended to establish rules for the creation of a market in the field of data and information

In accordance with decision No. 9 taken by the Kazak Cabinet of Ministers of on March 10, 1994, the Ministry of Ecology and Biological Resources of the Republic of Kazakhstan developed an integrated national system for ecological monitoring, which was to be coordinated among interested ministries. By a decision of the Prime Minister on January 3, 1995, the concept will be considered in conjunction with new wording to be given to the Law on Environmental Protection.

3.3. DRAWBACKS TO THE CURRENT SYSTEM OF MONITORING AND CONTROL

The present system of local departmental control over environmental affairs has several disadvantages:

- dissociation and methodical incompatibility of departmental services of ecological control, insufficiency of automation of processes of receiving, transmission, processing and propagation of information
- lack of a central network for processing and analyzing ecological information together with a lack of standardized methods for quantitative and quality measurements of resources and of anthropogenic impact, and for making assessments
- low quality of instrumentation and quality control of data; poor monitoring methods which produce poor data.

3.4. THE NATIONAL INTEGRATED SYSTEM OF ECOLOGICAL MONITORING

The purpose of the integrated system is the provision of required environmental data in various formats (digital, text, image cartographic, etc.) for all concerned government administrative units. In order to achieve this aim the integrated system would have to perform the following functions:

- Monitor environmental conditions and concomitant anthropogenic impacts, taking into account reactions of the biosphere and changes in sanitary conditions
- Undertake the systematic collection, processing, storage, and distribution of ecological data
- Assess the actual state of natural ecosystems with a view to discovering critical situations
- Prepare reports and develops short and long-term forecasts
- Prepare recommendations for improving environmental conditions in specific locales.

It is of a particular importance that the operations of both current and newly designed information and data networks be coordinated among Kazgidromet, Ministry of Ecology and Biological Resources, the National Academy of Sciences, Goskomzem, and the Sanitary Epidemiological Medical Service. The functioning of the integrated system has a sound legal basis. It should be allowed that the users of the system may be both suppliers and owners of the data. The various levels of the system, from national to local should interact through specialized centers and laboratories that share data. Within an integrated and coordinated system, the various units such as the ministries, Kazgidromet, Goskomzem, the Water Committee, etc., should remain autonomous but share methods and technology.

Within this context, surface water monitoring should be basin-wide and involve automated water quality control stations, and the data accumulated should be transmitted to central databases.

3.5. THE INTERSTATE ECOLOGICAL MONITORING SYSTEM (IEMS) OF THE CIS

The Interstate Ecological Council of the Commonwealth of Independent States (CIS) established the IEMS for the following purposes: to create a data foundation for the management of the environment of the CIS countries, to formulate coordinated policies in regard to environmental monitoring, the cooperative collection and analysis of data, assess environmental conditions in the CIS countries, establish standardized methods and measurements of monitoring, to create an integrated and coordinated system of data collection and data sharing to assist policy makers and managers, and to undertake various cooperative projects and programs. The IEMS is intended to function at the interstate, national, regional and local levels.

The exchange of data among the national centers of the IEC members is based on no-cost access to data. Access is arranged by means bilateral and multi-lateral agreements. Exchange of data between the national centers of IEC member states is accomplished on the principle of free of charge access to the monitoring data received at the expense of state budget means. Access to the ecological information is possible on the basis of bilateral or multilateral agreements.

4. Conclusion

Despite an increase in the flow of Kazakhstan's principal rivers in 1993 and 1994 as compared to the long-term averages, and despite some decrease in industrial waste discharge, the overall quality of the Republic's water resources remained unsatisfactory. There was no improvement in 1995.

In 1995, underground water resources also continued to be exposed to a great deal of anthropogenic pollution from 450 sources of pollution that were identified.

The quality of drinking water supply did not improved in 1994 as compared to preceding years At least 8.6% of sample specimens failed to meet the established hygienic requirements.

The ecological conditions of the Aral Sea continued to worsen. In 1994, the water level in the sea was 36.6 abs.m, the surface area was 29,600 km^2, the volume of water was 223.8 km^3. The continued rise in the level of the Caspian Sea increased the ecological, economic, and social problems of the region.

In 1994 there was no significant change in the health of the population. The deterioration of the environment continued to have a great negative impact upon the reproductive rates, the health/death rates especially of children and the aged.

Among the improvements made in the environmental sphere in 1994 were:

- Strengthening of environmental legislation, the adoption of international standards with a view of preparing Kazakhstan for participating in the international system of environmental security.

- Programs for the adoption by industry to non-waste/low-waste, closed-cycle technologies, recycled water supply systems, the development of local technologies for water purification, and the utilization of industrial collector-drainage sewage.

- Reconstruction, expansion, and new construction of the water supply, sewage systems, purification facilities, and wastes utilization structures in cities and large centers of population.

The problem of supplying the country's population with clean drinking water has not only persisted, but has even worsened slightly.

The impact of environmental degradation on the population continued to worsen. In 1995, every fourth person suffered from respiratory disease; one half of the republic population suffers from malignant tumors; rate of tuberculosis in regions lacking clean drinking water increases; the birth rate continues to fall and mortality rate to increase all over the Republic.

In 1995, the legislative and methodological basis of resource management in Kazakhstan continued to improve.

Notes

1. The national report "On the state of the environment of the Republic of Kazakhstan in 1994" ME&BR RK. - Almaty, 1995. 122 pp.

2. The State report "Ecologic state of the environment of the Republic of Kazakhstan in 1995 and measures taken for its improvement" / ME&BR RK. - Almaty, 1996. - 129 pp.

3. Concept of the United State System of the ecological monitoring of the Republic of Kazakhstan (USSEM RK) /ME&BR RK, AS RK. - Almaty, 199. - 13 pp.

4. Decision of the VIIth Session of the Interstate Economic Council No. 4 as of October 24, 1996 "On agreement on cooperation in the field of ecological monitoring and integration into international monitoring systems" - 7 pp

5. Decision of the VIIth Session of the Interstate Economic Council No. 5 as of October 24, 1996 "On provision on the interstate system of ecological monitoring" - 5 pp.

6. Information on the activity of the Ministry of Ecology and Biological Resources of the Republic of Kazakhstan and ecological situation on the territory of the Republic in 1996. /ME&BR RK. - Almaty, 1996. - 34 pp.

7. Information ecological bulletin of the RK for 1996 /ME&BR RK, MPH RK, SA of Hydrometeo RK. - Almaty, 1997. -68 pp.

8. Information ecological bulletin of the RK for 1995 /ME&BR RK, MPH RK, SA of Hydrometeo RK. - Almaty, 1996. -143 pges

9. Catalogue of the Automated Data Bases of the Republic of Kazakhstan B.A.Kembaev, E.I.Granovsky, V.P.Borodin, L.N.Moskalenko, V.M.Neshchadim, M.I.Ranzina.- Almaty KazgosINTI, 1966.- 209 pps.

10. Bibliography of additions of domestic and foreign scientific and technical literature in problems of ecology /KazgosINTI: Almaty, 1995-1997.

ll. Journal of abstracts- KazgosINTI, series 1, 1994-1997.

12. Bulletin of R&D registration, series 3, 1994-1997.

13. Collection of R&D abstracts, series 3, 1994-1997.

NEW APPROACHES TO WATER QUALITY MANAGEMENT IN THE RIVER BASINS OF UKRAINE

ANATOLY V. GRITSENKO
Ukraine Scientific Research Institute of Ecological Problems, USRIEP
6 Bakulina Street
310166 Khariv UKRAINE

Abstract

The quality of the environment in any country depends greatly upon an integrated water resources management system supported by appropriate and effective legislation. Prior to the present, Ukraine functioned under the former USSR legislative and institutional management system. The Soviet system had certain disadvantages that impeded the processes of efficient sustainable water resources management. Consequently, efforts have been made to improve the efficiency of environmental management in Ukraine. A significant example of the new approaches that are being implemented, particularly in the field of environmental information management, is the Regional Environmental Information Management System (REMIS) which has been applied to the Lower Dnieper basin.

Key Words

assessment, environment, information, management, monitoring, REMIS, Ukraine, water quality

1. Current State of Water and Environmental Assessments

Water quality management involves several elements: an integrated system of measures for the on-going assessment of the current condition of water bodies, the formulation of long-term objectives for the maintenance of water quality and quantity together with the development of mechanisms to reach these objectives, the capacity to predict changes in water quality within the modeling period, assessment of any improvements in the state of water bodies as a result of corrective measures taken, and improving such measures if they prove to be ineffective or their application proves to be too costly. Additionally, the steady assessment of the quality of water bodies and the ability to predict changes over the long term, a continuous water monitoring system is required.

T. Naff (ed.),
Data Sharing for International Water Resource Management: Eastern Europe, Russia and the CIS, 121–125.

However, until now, no efficient water quality monitoring system for Ukraine has been developed. Rather, several different incompatible systems have been in operation. Various ministries have developed and applied their own specialized water monitoring models, using different data formats, sampling frequency requirements, lists of the monitored parameters and analytical methods. Because of a lack of integration, these models cannot function in unison.

Classic water quality monitoring incorporates the following components:

- monitoring the impact of various activities on the water
- monitoring and assessment of ambient water quality
- prediction of changes in ambient water quality caused by anthropogenic activities
- long-term water quality assessment.

Traditionally, only the first two processes have been a part of water quality monitoring in Ukraine. The State Committee for Hydrometeorology has established about 250 monitoring stations throughout the nation. The frequency of measurements at these stations varies from once a month to seven times a year.

The collection, processing, and reporting of statistical data on the quantity and contents of wastewater is the responsibility of the State Committee for Water Management. Regional sanitary stations are in charge of operating wastewater treatment facilities that function periodically and randomly. Random control of water quality, both at the sources of wastewater discharge and in water bodies themselves, is also exercised by the Ukrainian Ministry of Environmental Protection and Nuclear Safety. All of the monitoring data produced by these units are desegregated and up to the present no attempt has been made to unify or integrate them. Consequently, these data can be hardly used for purposes of decision-making and strategic planning. It is also worth mentioning that these data are classified and therefore public access to them is very limited.

2. REMIS As a Tool of Water Quality Management

The REMIS approach, which emphasizes uniform standards of data measurements, collection, processing, reporting, integration, and sharing, enables the achievement of collection, analysis and integration of separate information flows into the common system for effective water quality management. Moreover, regional environmental management information systems make it possible to increase monitoring capacities and to activate those monitoring functions which have heretofore been exercised only occasionally.

REMIS can be used as an important integrated management tool by the regional environmental protection authorities to ensure efficient multi-purpose use of accumulated data and information on water quality and pollution sources by various users in a user feedback regime. Introducing REMIS would enable the regional environmental authorities to use the unified distributed information for operational decision-making and strategic planning purposes and to improve significantly the efficiency of undertaken and planned measures.

REMIS is a flexible system, one which can be updated by means of incorporating a water quality simulation module for analysis of changes in the ambient water quality under different economic development scenarios. The model also enables analysts to make recommendations on operational water quality management.

REMIS can serve as a link connecting organizations executing various monitoring programs and regional environmental protection authorities responsible for water quality management. In order to ensure reliable operation of REMIS and to maintain up-to-date scientific and technical level of platform software, adequate scientific and methodological support is needed at all stages of its development, implementation and commissioning. REMIS, being a multi-disciplinary instrument of water quality management, incorporates the elements of monitoring, management and pure science.

3. Phases of REMIS Development

In the first phase of REMIS development, several elements must be specified. They are: the content of the data, their formats, and the technical details of data transfer, the design of the software tools and the methodological principles that will be employed.

The REMIS system incorporates a number of computing modules for ambient water quality assessment and classification of water pollution sources. In addition, other modules are included which operate on the basis of available information only. These latter have no direct links with the issues of improving of the process of raw data collection from different ministries and authorities, nor with those issues related to the institutional restructuring and expansion of responsibilities of the regional environmental protection authorities.

The objectives of the second phase are:

- more detailed processing of available information to be used for water quality modeling for various economic development scenarios
- analysis of potential changes in water pollution sources
- development of recommendations for water protection measures.

The third phase of REMIS development includes activities for the updating of existing monitoring databases and expanding the responsibilities of the regional environmental authorities in the field of environmental information management. The application of the REMIS System in phases to the Lower Dnieper River is illustrated below.

TABLE 1. Model of REMIS application to Lower Dnieper River.

Phase 1	Standardization of data collection and analysis and water quality assessment.
Phase 2	Creation of REMIS model.
Phase 3	Data collection, online sharing, operational quality management and early warming systems.

On-line access to the operational information on water quality and sources of pollution by the regional environmental authorities is urgently needed if they are to undertake timely, appropriate measures in the event of accidental changes in water bodies under their jurisdiction. To achieve this purpose, coordination of the development and operation of water quality management and data collection is required.

4. The Application of REMIS to Large River Basins

To ensure efficient water quality management in large river basins — such as the Dnipro river basin in Ukraine — it would be desirable to develop a common, integrated information system consisting of several linked regional environmental management information systems. This common system would provide easy, efficient access to all data and information available in the various regional databases. There are three possible configurations for establishing links among regional information management systems:

In the first configuration, the regional environmental management information systems exchange data and information for solving specific transboundary problems and to coordinate activities basinwide. In the second, the regional systems transfer information to a basinwide information center and receive needed information on request. In the third instance, the participants in REMIS create a central, basinwide information node that becomes part of REMIS.

5. The Development of a REMIS Network for the River Basins of Ukraine

REMIS, as a tool for establishing integrated information management, information distribution, and sharing, can be applied to other river basins of Ukraine, the Danube, Dniester, Seversky Donets, and the Southern and Western Bug.

Generally speaking, the exchange of data and information among REMIS projects among river basins where there are no systemic or other connections makes little sense from the point of view of practical water resources management. However, the exchange of experiences on approaches used could be useful.

In Ukraine, the Ministry of Environmental Protection and Nuclear Safety, as the governmental body in charge of strategic water resources management and early warning of environmental accidents, would be the most appropriate agency to take responsibility for the collection of data and information on the various river basins. In this scenario, the information accumulated by REMIS groups of the different river basins would be presented in a unified format to the Department of River Basin Management of the Ministry of Environment for summarizing and review. However, before nationwide REMIS approach could be put in place, certain institutional and legal problems would have to be overcome. Chief among the constraints are the legal issues involved in implementing REMIS, the relationship of the system with owners of raw information and data, and the financial responsibilities involved.

Should REMIS be officially adopted, the Ministry of Environment of Ukraine would be provided with a powerful information exchange network, which would empower the Ministry to improve significantly the efficiency of water quality management among the river basins of Ukraine.

HUNGARIAN NATIONAL DATABASE OF WATER AND WASTEWATER MANAGEMENT

MARCELL KNOLMAR
Budapest Technical University
H-1111 Budapest HUNGARY
knolmar@vcst.bme.hu

A. DELI
National Water Authority
HUNGARY

Abstract

The National Water Authority (OVF) operates the national database of water supply and wastewater management. The database includes all the data produced by a man-made water cycle — beginning with the acquisition of water, continuing with its purification, delivery, distribution, canalization, its use for sewage collection, and ending with the delivery of sewage to a designated body of water. The data encompasses the quality and quantity of water and wastewater, the technical parameters of treatment technologies, the delivery and collection systems, and the owners of the public works. The regional water authorities are responsible for supplying and supervising the data and are also able to receive data from the entire database.

This study is intended to describe the information system developed for managing the public water and wastewater data in Hungary. In the second part of the paper a case study is presented to prove the usefulness of this information system. This example demonstrates the capabilities of the information system, what kind of and how detailed data can be gained from the database for a nutrient balance analysis.

Key Words

balance, data, Hungary, information, quality, sewerage, treatment technologies, wastewater, watershed, Zala

T. Naff (ed.),
Data Sharing for International Water Resource Management: Eastern Europe, Russia and the CIS, 127–133.

1. Introduction

Over the past 25 years, Hungary's water supply and canalization systems in settled areas have undergone significant development. Between 1970 and 1990, public water works received most of the developmental attention in the form of an expenditure of U.S. $100 million, and since 1990 the focus has been on canalization. The end of communism in Hungary resulted in a multiplicity of the companies involved in public works. At the same time, greater responsibility for the public administration of water shifted to local governments whose role was thus altered fundamentally.

The new conditions lead to a situation where data associated with local governments increased both in importance and quantity. The demand for the management, overview, and utilization of data became more and more intensive at both the ministerial level and among regional water authorities.

Until the beginning of the present decade, separate information systems had been developed for the management of the growing amount of data such as registration data of the predetermined subsidy system and registration data of the industrial water uses management. However, these separate information systems became quickly outmoded in relation to needs and there emerged a demand for a unified system that could serve the requirements of settled regions. The wide availability of GIS software and the rapid decrease in the price of hardware opened the way for the use of GIS in Hungary.

2. The Public Water Works Information System

In the public administration of water, the two main sources of statistical data acquisition concerning water infrastructure are:

- statistics of urban based public water works of the Central Statistics Office (CSO)
- statistics developed by the water administration sector itself.

Obviously, urban areas are the prime focus of the CSO. It is from these centers that most of the data concerning public water works are collected. The companies that run public water works provide the most typical data from their day-to-day operations — for example, the number of wastewater treatment plants, the length of sewer networks operated by a company, the distribution of the sewers according to diameter, age, etc.

The Water Ministry (KHVM) initiated the development of an information system in 1992 using pre-existing data units. The project was carried out by the Water Research Center (VITUKI) financed from research and development funds and in 1994 the new information system began to function. The leading Hungarian GIS company (GEOVIEW) developed the computer technology dimension of the system. The system operates on two different platforms: one in the KHVM uses a SUN workstation under a UNIX operating system within the ORACLE and ARC/INFO software environment, supported by a special program developed by VITUKI; the other in the National Water

Authority (OVF) on PCs using the Windows operating system in an ArcView environment without any special programs.

Both systems use the same database, employing maps at a 1:500,000 scale, the most important layer containing data on the administratively independent settlements in polygons, lines, and points.

The textual data used by the information system is as follows:

General data on the settlements

- name of settlements
- name of county and regional water authority
- number of permanent inhabitants, number of dwellings, size of administrative area
- surface and underground water quality category of the settlement
- new water acquisition protection category

Data on development

- data on subsidies for developing public water works
- purpose, dates, and amount of subsidy
- results of investments carried out
- water sector data
- government investments (based on approval documents)

Technical data on water supply and sewerage for settlements (based on KSH statistics)

- water production
- water service
- dwellings connected to water supply network
- water supply network
- water supply connections to homes
- temporary water supply
- dwellings connected to sewer network
- wastewater treatment

Data on water supply and sewer network operating companies in settled areas

- basic data on operating companies
- tariffs

The current information system has certain limitations:

- The basic lines of administrative authority for data in independent settlements are not clear, especially in regards to those settlements that were formerly independent (and in certain structural ways remain so) but have now been administratively incorporated into larger cities.

- Owing to the way that urban centers are administratively structured, the regional connections between water companies cannot be registered

- Regional water authorities exhibit a low level of interest in the present information system because the system fails to give adequate support at regional levels.

The conclusion to be drawn from these limitations is that the statistical data method could not be employed beyond the county level, leaving the regional level largely without coverage. This situation has made changes in the system unavoidable.

3. Changes in the Public Water Works Information System

The most important innovation introduced into the statistical system for public water works is that responsibility remains in the public rather than private sector (i.e. the operating water companies). This means that data are more constant and long-term. [1] An additional advantage is that the statistical data can be used in conjunction with maps where their sources can be located, making them more compatible with the use of GIS.

In any case, a given water work or wastewater system can be separated into units. Under the new statistical system, these units may be classified in the following way:

Water Supply

- water production and purification (surface and ground water intakes and treatment)
- water supply (i.e., delivery, not distribution)
- water distribution (i.e., piped drinking water to consumers via dwelling connections).

Wastewater Disposal

- wastewater collection networks (only for dwellings with piped water)
- wastewater delivery (only for long distances to treatment sites; excludes connected dwellings)
- wastewater treatment plants, including sludge treatment technology.

These systems function reciprocally, receiving water or wastewater from one another. In some instances, the two systems are interconnected. The water transfer can be an operation by the same company or from a foreign partner company as well. Not all of the aforementioned functions exist in all the units of the system. In some, water delivery is absent because the water flows directly from the site of production to the distribution system. These activities and systems are embedded in the changed statistical data methods, meaning that these activities are supposed to be prepared on data sheets which are edited specifically for each kind of function.

4. The New Water Database: Zala Case Study

During the past three decades, the consequences of high nutrient loads have had an adverse effect on the quality of Danube and Black Sea water. Emissions of phosphorus and nitrogen were detected in all parts of the antroposphere. High nutrient loads and their consequences were recognized as a severe water quality problem of the Danube and its receiving water body, the Black Sea, during the past decades. High emissions of phosphorus and nitrogen were detected throughout the antroposhpere. To develop a control strategy for those emissions, a detailed analysis of the nutrient metabolism of the antroposphere was necessary.

A nutrient balance study was undertaken to analyze and evaluate the N and P turnover which was driven by Hungarian economic development and by natural resources. [2] The case study was aimed at identifying the determinants of N and P paths and the quantification of fluxes and flows from the antroposphere into the environment, with special attention to the nutrient discharges of Hungarian surface waters.

The scope of study included point-like and diffuse nitrogen and phosphorus emissions originating from different economic sectors, emissions reaching surface and groundwater, and the governing processes in the antroposphere. Among the tasks performed were an evaluation of the present level of the water supply, sewerage, wastewater treatment and sludge disposal, an assessment of major strategic problems, and the preparation of a foundation for future strategy development.

Information and data were available from the reports of the Central Statistics Office, regional (county) statistical offices, the National Water Authority, the regional water authorities, regional environmental inspectorates, and water and wastewater works operating in the region. Climatic data were supplied by the National Meteorological Service and by published statistical annuals.

Information on water supply and wastewater treatment conditions in Hungary is rather detailed. While 81 percent of the population lives in dwellings connected to a public water supply network, only 51 percent have public sewerage. The utility gap is very large. Moreover, only 20 percent of the population living in unsewered areas is serviced by adequate local waste disposal methods. Household sources account for about 40 percent of total waste, industry for 25 percent, institutions for 12 percent, and storm water together with groundwater infiltration contribute 20 percent. There is a striking pattern in the level of wastewater technology: 12 percent of municipal sewage receives only mechanical treatment and 31 percent receives biological treatment, but 55 percent receives no treatment whatsoever. However, despite the relatively low percentage of treated wastewater, the overall quality of Hungarian surface water is good, because of the relatively high dilution rate.

A fairly detailed evaluation of nutrient balance was carried out in the catchment area of the Zala River located in the southwestern part of Hungary. This area forms the western segment of Lake Balaton watershed. The length of the watershed is 2700 km^2, mostly in Zala County. It is one of the very few watersheds that lies entirely within Hungary. Water quality data for the Zala River are the most detailed in Hungary.

Plains, mud flats, and hills characterize the watershed. Human activity is the most important factor in modifying the landscape and ground surface. The watershed is an agricultural area with a low population density.

The water supply in the Zala watershed has been developed significantly. Presently, 92 percent of the dwellings are supplied with water and 94 percent of all water produced is distributed. Of the distributed water, 73 percent is consumed by the population and the rest serves other consumers such as agriculture and industry. Data from regional water works show that both industrial and domestic water consumption has decreased dramatically in the recent years. In this situation, establishing public utilities at regional and subregional levels would appear to be the most sensible policy solution. The establishment of public water supplies accelerated in the 1970s and 1980s, but, unfortunately, here again the sewerage did not keep up with the rapid development of water supplies.

It is only in the last 15 years that a significant improvement of canalization and wastewater treatment occurred. In 1981, only nine towns had canalization and some of the seven water treatment plants located in the area were overloaded, had capacity problems, and neither biological treatment nor nutrient removal was being performed anywhere. However, in the early 1990s intensive development of new sewer systems and treatment plants was implemented

Currently, there are a total of 63,755 dwellings in the cast study area. Ninety-two percent are connected to a water supply although only 39 percent of them are connected to a sewerage system. The total length of the sewerage system is 319 km, mostly separated (197 km). According to the most recent surveys, in Zalaegerszeg, the largest town of the Zala catchment, the level of sewerage connection is 84 percent. The relatively high canalization data do not reflect the fact that the number of inhabitants connected to sewerage is relatively much lower. Local governments are making serious efforts to increase the rate of connections. In the catchment area, there is no settlement with a sewerage system without sewage treatment, so that the amount of untreated sewage transported by sewer system is zero. The sewage that is produced is collected in septic tanks in most villages. There are sewage treatment plants in the following settlements: Zalaegerszeg, Zalaszentgrot, Zalakaros, Zalakomar, Zalalovo, Zalaapati, Turje, Hahot, Sarmellek, Varvolgy, Zalacsany, Gersekarat and Hegyhatszentjakab. The wastewater is treated mechanically and biologically, and there is tertiary treatment with chemical P precipitation at almost of these plants. The amount of treated wastewater is 16,349 m^3 per day and most of this is domestic wastewater.

A regional canalization system is being established in the region of Zalaegerszeg. By this system, the wastewater of seven villages in the area is collected and directed to Zalaegerszeg. Due to changes of the governmental funding system, there are now few new investments with the consequence that the rate of improving public canalization has slowed significantly.

Of course, the negative effects on the area of the quick growth of the canalization system, between 1991 and 1994, were also detectable. Often, owing to a combination of shortages in time for preparations and coordination and to a greater assertion of independence by local governments, the most optimal solutions were often not chosen. On many occasions, even habitations that were forming conglomerates could not be convinced of the advantages of establishing a common canal system or water treatment plant. Too many small, individual plants were established, most of which are inefficiently used and have high costs. For these reasons, particularly in the first part of this decade, even the start-up operations of treatment plants caused problems. One plant, for example, was not used for two years after its construction because of insufficient wastewater. In the meantime, connecting users to the public service drags on.

References

1. Deli, A. and L. Szabo (1996) *Megvaltozott az agazati vizikozmu statisztikai rendszer* (The Statistical System of the Sectoral Public Works has been Changed), Viztukor (Az Orszagos Vizugyi Foigazgatosag lapja).

2. Buzas, K. (1996) *Nutrient Balances for Danube Countries*, Technical Report, Danube Applied Research Program (PHARE: ZZ9111/0102).

DATABASES AND INFORMATION SYSTEMS (IS) OF THE ARAL SEA BASIN STATES

V. P. KROHMAL
Chief Engineer Deputy
Institute "Turkmengiprovodhoz", TURKMENISTAN

A. A. VECHER
Chief of the Computer Center
Institute "Turkmengiprovodhoz", TURKMENISTAN

Abstract

Until 1991 water resource planning and use in the Central Asian Republics (CAR — Kazakhstan, Kyrgyzstan, Tadjikistan, Turkmenistan, and Uzbekistan), particularly as regards the Amu-Darya and Sir-Darya Rivers, were carried out in accordance with the priorities and long-term plans of the former Soviet Union without regard to the interests and needs of the CAR. Soviet policies created very serious ecological and water problems in the Aral Sea region. Solutions to those problems requires the establishment of a distributed database network, cooperative data collection and data sharing among the members of the CAR. Various technical and budgetary problems make the achievement of this goal difficult. This study discusses what is being done to overcome the obstacles.

Key Words

Amu-Darya, Aral Sea, CAR, database, DIS, ecology, GIDROMET, information, Sir-Darya, Turkmenistan, WARMAP, WARMIS

1. Introduction

Given the general arid conditions in the region of the CAR, water plays a vital role in the economic and social development and in the ecology of all the member nations. Until 1991 water resource planning and use in the Central Asian Republics (Kazakhstan, Kyrgyzstan, Tadjikistan, Turkmenistan, and Uzbekistan), particularly with regards to the economic exploitation of the Amu-Darya and Sir-Darya Rivers, were carried out in accordance with the priorities and long-term plans of the former Soviet Ministry of Water and Reclamation without regard to the interests and needs of the

T. Naff (ed.),
Data Sharing for International Water Resource Management: Eastern Europe, Russia and the CIS, 135–144.

CAR. Soviet policy over many decades emphasized large-scale, intensive irrigation programs which created very serious ecological and water problems in the Aral Sea region.

The Aral Sea, which is a closed basin that receives the flow of Central Asia's two largest rivers, the Amu-Darya and Sir-Darya, encompasses approximately 690,000 km^2. These rivers receive directly the industrial and municipal wastes of most urban and rural settlements in the region, which constitute the main source of the rivers' contamination. Another basic source of pollution of the two rivers is the drainage water from irrigated fields which contain high levels of harmful chemicals and salts. All of these pollutants, in very large quantities, have for years been dumped into the Aral Sea.

As a result of these harmful practices and a large population growth, the Aral Sea had lost 37,000 km^2 of its area and had fallen by about 16 meters (m) by 1995. In some places the Aral Sea coast has receded up to 120 km from its original position and the level of salination has increased by a factor of three. The results have been a very sharp deterioration of water quality in the Sea, serious adverse effects on public health, on fish reserves and the fishing industry as well as a reduction in agricultural efficiency and production.

Following the end of the Soviet Union and the creation of the independent states that constitute the CAR, new political and economic conditions have made possible the formulation of new, more ecologically and environmentally rational water resource policies, especially in regard to the Aral Sea. The CAR members of the Aral Sea basin are in a position to develop cooperative strategic water policies regarding the sharing and use of the Aral Sea.

However, from the very inception of independence from the Soviet Union, there has been little exchange of water-related information and data among the nations of the CAR. Too few official channels for such exchanges exist. Without cooperative collection and sharing of data it is not possible to carry out effective policies for the development of surface and ground water resources in the Aral Sea region. Fortunately, the heads of state of the CAR have recently agreed to cooperate in the joint management and distribution of water facilities, to create an interstate management organization (BVO), and undertake other coordinated activities regarding the Amu-Darya and Sir-Darya Rivers.

It is necessary to point out, however, that despite good intentions for improving the water and ecological situation in the region, the members of the CAR are burdened with the legacy of the Soviet Union, including a syndrome of very serious, unmitigated problems and with very limited resources to deal with them. In order to implement the agreement of cooperation in dealing with the region's water problems, the CAR members will need considerable financial and expert assistance from the international community. Toward this end, in 1992 the five member nations of the CAR requested aid from the European Union to improve surface and ground water resources and agricultural production. In January 1994 an aid package for agro-economic development and the improvement of the ecology of the Aral Sea region, funded by the World Bank and the EU, was agreed upon with the five CAR members. The project was labeled WARMAP — Water Resources Management and Agricultural Production.

Under the original plan, nine projects were to be undertaken. But, owing to a shortage of funds during the planning phase (January to August 1995), the number was reduced to seven. In the implementation phase (September 1995 to July 1997) the international and local organizations involved focused on six areas of activity:

- The preparation of the necessary interstate agreements, which would ensure the management of water resources management on a regional basis.
- The creation of a regional data, information and communication system with specialized units for the efficient management of the region's water resources.
- An analysis of irrigation water use and its efficient management.
- A pilot project for the improvement of irrigation engineering.
- A pilot project designed to improve the administration and maintenance of interstate economic channels of information.
- The evaluation of water quality and ways to improve the quality of agricultural water.
- Purifying water and improving public health.

2. Problems of Database Systems in the Aral Sea Water Improvement Project

Obvioulsy, program efficiency, justification of costs, and assured results and profits from the Aral Sea improvement project are determined by many factors, such as the comprehensiveness and quality of the available data, the organization and reliablity of communications linkages among participating partners, the establishment of a workable interstate data system which involves integrated institutional structures, a rigorous system of monitoring both the technical and programatic aspects of the project, and, in general, ensuring that the political, financial, organizational, methodological dimensions of the issues are properly attended to, and, of course, ensuring that data and information are open and available.

Although there are large, expensive data and information systems in other parts of the world, the one proposed for the Aral Sea project is different in both scale and purpose. An analysis of the process planning and designing proposed information system between 1995 and 1997 can be informative to experts within and outside the region, especially to any who would like to participate in the creation and maintenance of the system.

Prior to the commencement of the project, information and data on the development, recording, and use of ground water resources were not exchanged among the five countries in the Aral Sea basin. Within these countries, available information was dispersed among various departments and ministries. Presently, ground water resources data are difficult to obtain because of barriers put up by the ministries and departments, or for reasons of confidentiality, and now for commercial reasons too. Information has become a highly saleable commodity for purposes of income.

In addition there are several technical problems that present added obstacles to free the accession of data. In the late 1980s and early 1990s an effort was made in various ministrics to create an interstate database. However, logistical and financial problems combined with planning that centered on an outmoded USECM type mainframe computer stifled the endeavor. Moreover, due partly to the obsolete computer and the delays involved, effort was viewed by many as experimental and not a basic tool for their work, so neither time nor a place was made for such a database in the various organizations it was intended to serve.

When desk-top computers were widely introduced following the demise of the Soviet Union, efforts to create a large, integrated database were quickly terminated and the data that was collected on punched computer cards have been either unused or lost. The use of the desk-top computers to network data was severely limited owing to the lack of appropriate software. Presently, the huge, non-systematic accumulation of useful data is not recorded and stored electronically, but on paper.

The information and data collected by various ministries and departments is neither categorized nor integrated, making it necessary to search for needed data in several departments. Compounding the problems are the frequent, often large-scale reorganizations of ministries and departments. During this process, data and information is constantly under threat of loss. The recent regional economic recession over the last several years has resulted in a serious deterioration of both the quantity and quality of collected data, including that which is necessary for the planning and management of water-related activities.

The data and information infrastructure in the Aral Sea region does not conform to modern requirements and thus does not have any capacity for the remote collection of data. The absence of an efficient and open system of information exchange among the CAR members and even among the relevant agencies within each state aggravates all the water-related data problems.

Until recently, the principle source of technical and other kinds of water information has been GIDROMET of the World Meteorological Organization. It was assumed by World Bank and EU experts, in the early stages of actions taken on the Aral Sea problems, that they would be able to receive on-line information from WMO's GIDROMET. They hoped that hydrometeorologic data for monitoring the natural environment in the Aral Sea basin area would be available and that it would thus be possible to analyze and make hydrological prognoses which would identify various technical problems and managerial problems and could underpin the development of better planning and policy making. In 1995, the WARMAP project attempted to fulfill these objectives.

It became quickly obvious that the data required to deal with the very difficult and complex problems of the Aral Sea — at least whatever data exists — is in the hands of the water resources agencies of the CAR.

3. The Development of a CAR Information and Database System

The initial feasibility studies — which involved both international and CAR experts — for the establishment of a CAR information and database system were undertaken under the auspices of the World Bank project for the creation of "a regional unified information system for the registration and use of water resources and hydro-ecological monitoring." The main purposes of the project were stipulated as:

- Creating an electronic water resources database and hydro-ecological monitoring system at both national and regional levels
- Making accurate information open and available
- Developing mathematical models to evaluate and predict water resources use for the improvement of water quality and water-related decision making in the region
- Creating an integrated system of water data record keeping and management and improving data collection and services
- Upgrading the maintenance of data and information for all projects particularly those related to the improving the ecology of the Aral Sea.

The new information system would consist of three main components: a geographical information system (GIS), an environmental monitoring system, and an information management system. The information system would function at five related levels: the basin, the republics, the region, specific areas, and specific fields. Data and information would be collected and shared both vertically and horizontally and would be integrated in the database. Organization and administration would reflect this design.

There would be three interstate operational units — the ICWC Research Center, BVO Amu-Darya, and BVO Sir-Darya. The results of the research and other activities in these units would be transmitted to all organizations and agencies and water users who need them. The data will become a part of GIDROMET through observation stations in the CAR countries.

This would be a distributed database system in all five Central Asian republics at all five levels of operation. Monitoring and analyses of ecological, hydrological, and environmental conditions would cover surface and ground water resources and the Aral Sea basin. As planned, the system was designed to resemble a star-shaped structure that would link and serve between 10 and 20 database servers and local networks. This polygon model would be tested in each republic on one experiment where the methods of collection, processing, and transfer of data from one level to the next on such issues as salinity and mineral and chemical content in ground and surface waters.

In accordance with the organization plan, the core of the distributed database system would be a regional coordinating center (the ICWC Center) where the financial and material resources would be concentrated. All data would flow to this regional center which would then process them and make them available on-line to all the users in the system, principally the CAR national centers. The center would not only coordinate but keep the data current. The plan assumes that once the system is organized and functional, any partners — i.e., the initial suppliers of data and information — could be immediately connected to the core regional center which would be situated in Uzbekistan. The system would impose rigid requirements on all participating countries to ensure the flow of information on any projects that concern the Aral Sea problem.

4. Assessment of the Proposed Data and Information System (DIS)

A cardinal weakness of the proposed plan is that the core center concept would locate a regional center in one of the republics of the CAR (Uzbekistan) and relegate the others to the status of data suppliers and the recipients of "off-the-shelf" solutions. There appears to be hierarchy in the management, authority, and functions of the system. That approach does not honor the principle of equality and independent decision making among the Aral Sea basin nations. The organization and methods employed appear to be excessively complicated and confusing.

The central system would be more expensive and require more equipment, software, maintenance, and trained staff than separate local, i.e., national systems which could coordinate with one another. Any regional system should be based on high speed, quality communications which are already developed and use telephone lines and have e-mail capability.

The differences in technical terminolgy, methods of measurements and analyses among the users could create data problems that the central server could not overcome, particularly in situations of potential conflict. The participants in the system would probably insist on limiting and regulating access to data.

Other unclarified issues include the terms of reference under which the DIS would be created, the nomenclature that would be used, the purview of its functions and authority and the range of problems on which data would be collected. The only thing known with certainty is that the equipment that would be used is outdated and would not have the capacity for automated monitoring. It is likely that the proposed DIS would be outmoded even before it is put in place.

The reliability and safety of the proposed DIS, the quality of its information service, and the absence of national database centers in all the member states would be unacceptable. The limited resources and very large number of environmental and water resource problems in the CAR countries must compete for budgets with other national sectors such as military spending and energy. This means that assistance for the creation of the proposed environmental information systems from the international financial community would be, of necessity, quite considerable, but such a flawed plan is not likely to be an attractive investment.

5. Turkmengiprovodhoz Institute (TI) Information System Plan

The Institute's plan would avoid the problems of the proposed DIS by organizing a system that would be based on local data networks and national data centers which would have coordinating, managerial, and up-dating responsibilities. This plan would allow autonomy for the national data centers which would generate databased on particular kinds of problems identified by each national center in the program.

The unity of the information system would not depend on distributed database network that would require the collection and maintenance in each member database an identical set of data. Rather, under the TI plan, each autonomous database in the network would give to the other national data centers assured access to agreed upon categories of needed data in which they have a mutual interest. Formats, processes, operations and help information would be standardized.

This approach conforms more closely to the political realities among the CAR and would require smaller financial investments and fewer technical and human resources than the proposed DIS plan.

6. Requirements of the TI Plan

The implementation of the TI plan would entail several steps:

- Normalizing and coordinating an electronic system of information exchanges among the national data centers.
- Provisions by various national and regional centers of data collected over the past several years, especially for Turkmenistan because the receipt of information and scientific and technical literature has virtually ceased since 1992. Additionally, an information program would have to be designed for organizing and processing the new (and backlogged) information and data received.
- An appropriate technical program for establishing and operating the entire system should be selected by the system participants and set up by experts. The design of the system should allow the flexibility necessary for each autonomous national center to adapt it to its own particular needs while maintaining standardized nomenclature and formats and periodic updates. The information exchanged should be only that necessary for the creation of a regional database.
- The acquisition of the most cost-effective, user-friendly, low maintenance, powerful, multiplatform, easily-upgradable hardware and software possible.

Taking these steps into account, it should be added that whatever information system is installed at whatever level — and, clearly, international funding and expertise will be required — it should be designed specifically with the problems of the Aral Sea in mind. Local experts should, at all stages be involved with their foreign counterparts if for no other reason than for the excellent learning experience they would gain.

It should be noted that prior to 1995 a compromise version of the original DIS was agreed upon by CAR experts after several meetings. A special working group was created to deal with any special problems that might arise. However, owing to a lack of funds, no action was taken on initiating the project.

7. Efforts to Create a DIS Since 1995

In 1995, under the WARMAP project, funded by the World Bank and the EU, a regional information system on surface and ground water among the five CAR states was undertaken. The initial version — Water Resources Management Information System, which was dubbed WARMIS-1 — was prepared by specialists of the WARMAP project. However, the effort foundered because the planners failed to take sufficiently into account the political and technical realities of the region. After several months of chaotic data gathering, experts in the ICWC Research Center and the TI prepared a WARMIS-2 (W-2) plan which was offered in February 1996.

After consultations with the relevant authorities in the CAR members, WARMIS-2 was formally proposed under the original WARMAP project. W-2 would have three basic components: Text databases in the Microsoft software environment; a GIS with appropriate graphic databases; and a system for sustained information exchanges.

Within this frame, the W-2 information system would do the following:

- Collect, process, and store data and information on surface and groundwaters, climate, the ecology, and socio-economic activities
- Exchange data and information among all participating members of the WARMIS-2 project
- Collect the necessary data for a wide range of analytical models for the purpose of solving water-related problems, for developing effective planning strategies among the CAR countries, and for training local staff
- Focus on data within the boundaries of the CAR and the associated region
- Collect data on the principal river systems, reservoirs, hydropower installations, and irrigation and drainage systems,
- Collect maps on underground water deposits including maps that indicate the depth and mineralization
- Collect soil maps including those that indicate the degree of salinity.

It is assumed that GIS topographical maps and special theme maps would require satellite photos that would be accessed by means of geocodes. Two other elements would be of particular importance: the W-2 database system should function in Russian and English, and it should have the capacity for linkages between text and graphics which would allow for the visualization and printing of data as reports, diagrams, graphs, etc.

The W-2 system would function primarily at the regional and national levels but also would receive information from WMO's GIDROMET. From the latter it would receive information on water resources and climate and would take local, i.e., national and regional, data on water resource uses and agricultural conditions from the various ministries and agencies.

Unlike the core regional data center that was part of the original WARMAP project, by which all data with standardized categories and formats would flow from the national centers and other agencies to the core unit, under the W-2 model there would be far less regulation; each autonomous national center would determine to a far greater degree what data would be shared, presumably on a need basis. W-2 would require precise statements of problems and needs, coordination among the separate units of the system, agreed-upon nomenclature, and the collection and transfer of data would be dependent on the availability of funds.

8. A New Information System Program

WARMAP has agreed to fund the development and operation of a new system. Instead of a single regional core IS center, each national IS center will function as a core center in its own right. Any regional centers would function in the same way. Experts provided by WARMAP together with counterparts from the ICWC Research Center and other national units designed a computer program for an experimental W-2 database. Water-related data collected from the period 1986 -1995, covering issues of surface and ground water, climate, quality, hydro-power, industry, drainage system, economic development, etc., was put into the database.

GIS will be employed to collect data and information on such issues as irrigation, climate zones, resevoirs, lakes, large rivers, channels, etc. All items will be codified and be translatable into text. The GIS maps will be on a scale of 1:500,000 using a Gauss-Knuger projection. The national groups will prepare special digitized maps on various topics, e.g., soil, salinity, ground water levels, etc. This project will be executed as soon as the necessary GIS equipment, which has been paid for by the WARMAP project, is received.

For purposes of analyses, the data will be modularized. The system is undergoing tests on two modules for water flow and salinity in the national centers. Upon successful completion of the tests, a whole series of modules such as hydro-power or economic development will be created. This approach would allow more effective use of the W-2 database. Particular attention will be given to the development of simulation and optimization models for the Aral Sea basin with a view to developing strategies for the rational use, protection, and distribution of the water.

Although a shortage of funds has prevented the installation of Internet linkages among the W-2 participants, nevertheless, progress is being made: data is being accumulated and tabulated. Plans for 1998 included increasing and updating water, water use, and economic data; creating sub-bases on hydro-power, ecology, and the Aral Sea; improving the response to inquiries and access to the data and reports at both the national and regional levels; improving the verification of data and information submitted by national groups; undertaking a study for the coordination of W-2 with other data networks; add data up to 1997.

9. Problems of Policy

In the course of developing W-2, differences between WARMAP authorities and CAR experts arose over certain matters of policy that have raised fears that the World Bank and EU might end their funding and the project would not be completed. The main issue of contention is the question of how openly accessible the data would be.

The WARMAP experts have taken the position that since the project is being funded by the international financial community, the data should be accessible without restrictions internationally. CAR officials have taken the stand that there should be only limited access to the data for several reasons: the data has national security implications and therefore, some of it needs to be classified in accordance with national statutes; some of the data would reveal national commercial trade secrets; some data falls under copyright; given the chronic financial straits of the CAR, the owners of the data have a legal right to recover the costs of collecting and managing the data.

10. Short-Term Needs

All the problems manifest in the effort to establish an effective W-2 system of data and information exchange among the countries of the CAR make clear the necessity for a comprehensive agreement among all the participants in the network.

In 1996-97 a sub-project, called WUFMAS, was successfully undertaken within the framework of WARMAP. A study of water use and management on representative farms in all five of the CAR countries was conducted. Data was gathered on planting, fertilizing, the use of chemicals, irrigation, and various other agricultural activities, including management practices were analyzed with a view to developing a set of recommendations that would make surface and ground water use more efficient in agriculture. In addition to the recommendations, the project provided an extremely valuable profile which will allow for accurate water use and agricultural production. Included in this pilot project was the creation of a database into which the collected data was put and can be shared by all the participants. In terms of data and information, the importance of WUFMAS is that it provides a model for data collection at a fourth level, i.e., directly on-farm.

What is needed for the foreseeable future, most urgently in the short term, is continued funding by the EU and other international financial institutions not only for the completion of the multilayered information system that has begun — which will necessitate provision of the essential technology and training — but to assist with other ancillary problems that have manifested themselves in the course of the WARMAP and W-2 projects. The next phase in the development of an information system will be to coordinate the development of a regional information system consisting of autonomous national databases which will gather data from the four levels of water-related activities: regional, national, local, and on-farm.

11. Sources of Information on the WARMAP Projects

The best sources of information on these projects, particularly with regards to the Aral Sea are the Reports of the ICWC Research Center and the *Bulletin of the Aral Sea* issued by the EC-ICAS. The widest range of information is contained in the reports issued by the national and regional centers, and by the WARMAP Project reports themselves.

A DATABASE ON ENVIRONMENTAL IMPACT ASSESSMENT IN A TRANSBOUNDARY CONTEXT (EIATC)

ANNA KWIATKOWSKA AND ANDRZEJ KRASZEWSKI
Warsaw University of Technology
Institute of Environmental Engineering Systems
Nowowiejska 20 00-653, Warsaw, POLAND
anka@iispw.edu.pl; aak@iispw.edu.pl

Abstract

This study describes the design, organization, and uses of a new environmental impact assessment database created under the auspices of the 1991 Espoo Convention. The database is designed for worldwide use on the Internet.

Key Words

database, EIA, EIATC, environmental impact, Espoo Convention, information, Internet, WWW

1. Introduction

The Convention on Environmental Impact Assessment in a Transboundary Context (EIATC) was signed on 25 February 1991 in Espoo, Finland. It applies to all project proposals that may have an impact on the environment of neighboring countries and which therefore are subject to a special procedure defined by the Convention. By this procedure a party who might be affected by the project in question must be given adequate advance information and has the right to influence the implementation of the project to avoid harm to itself.

A basic instrument for defining a potential adverse impact is the Environmental Impact Assessment (EIA), which is a multistage procedure implemented under the regulations of the country of origin of the project. All OECD countries have already gained significant experience in EIA procedures and even some practical know-how related to international cooperation in this field. The countries of Central and Eastern Europe however, are sometimes only at the developmental stage of such procedures. It seems therefore particularly important to create instruments that will facilitate through the exchange of information both the development of their own procedures and their adjustment to the international requirements of EIA as defined by the Convention.

T. Naff (ed.),
Data Sharing for International Water Resource Management: Eastern Europe, Russia and the CIS, 145–151.

Key to this system is a database that will collect information concerning project proposals that fall within the purview of the Convention. Such a database will provide interested parties with data on the implementation of EIA for projects, enable the storage of data on past EIAs, and contain information on the relevant legal regulations in the countries that subscribe to the Espoo Convention. The database structure was designed to facilitate implementation of the Convention to the greatest possible extent.

Environmental Impact Assessment in a Transboundary Context (EIATC) is a database designed for world-wide use on the Internet. Users who employ the database's network protocols, are provided with the following features:

- Accessibility through a user-friendly interface based on the World Wide Web (WWW) system. For users who do not have not convenient access to the Web (e.g., in places with low-band Internet lines), an e-mail gateway will allow access and modification of data. Users who have no access to Internet at all may acquire entree by having an MS-Access system installed on their personal computers.

- Information from EIATC will always be up-to-date and new records available immediately upon being entered into the database.

- An unlimited number of users will be able to browse simultaneously through the database contents, modify and add new records (except for the MS-ACCESS system) in compliance with data access rights.

- Parts of the database contents and modifying procedures will be protected and made accessible only to authorized user groups.

- Full on-line help will be available to users employing all three types of access, describing in detail all the database operations.

- Users may at any time send comments about the database and its contents to the EIATC operator and to authors of any documents in the database. Such comments constitute an important knowledge pool, useful for future upgrades of the database.

- All functions of the database, including help and information, will be available in English or Russian and it will be possible to make the choice while operating the system.

The EIATC database was designed at the request the UNEP/WHO Project Office, division of Environmental Impact Assessment at the Ministry of Environmental Protection, Natural Resources and Forestry. The design was prepared jointly by the Institute of Environmental Engineering Systems of Warsaw University of Technology and the WWW Technology Company. Once the project is completed, EIATC should become a dynamic information system encompassing users in many countries, and ensuring a broad exchange of expertise and knowledge among interested parties.

2. Overall Functional Description of the Database

The EIATC database, which will be accessible through WWW, will have a client-server type of architecture. The server will be located in the UN/ECE Secretariat in Geneva, but some of the supervisory functions can be transferred to the International Centre for EIA in a Transboundary Context situated Warsaw. Any authorized organization or institution in any country can become a client of the database.

The EIATC database will contain data that originates in the signatory nations of the Espoo Convention, and will concern a broad spectrum of EIA-related issues for all projects that may have a significant transboundary importance. Apart from typical textual and quantitative data, the database will be able to offer information of almost any type thanks to the use of the WWW multimedia system. EIATC will hold pictures, maps, and technical drawings in any electronic format (PostScript, GIF, TIFF, etc.). Voice data and video sequences may also be attached.

EIATC server software will provide some important additional functions:

- system access statistics (data collection, analysis, presentation)
- interface for its administrators either directly through a server console or in an agreed upon restricted form through the WWW system and/or electronic mail
- a system for discussions among database users — through the "mailing list manager" — in which discussion archives may be accessed through e-mail and WWW systems.

3. Database Users

As indicated, the database is designed for all institutions and organizations interested in EIA for any project that falls under the Espoo Convention. Within that framework, the following user conditions and functions would apply:

- *The Competent Authority of the Interested Party* (Focal Point for the implementation of the UN/ECE Convention). In accordance with the Convention, the interested party is responsible for notifying the competent authority of affected party (Point of Contact for Notification) about a proposed activity (project) and for providing relevant information. The affected party would be responsible for entering the project information into the database and its subsequent updating during the EIA process.
- *Competent Authority of Affected Party.* Would oversee the EIA process and would be authorized to submit reservations concerning the procedure. Apart from the exchange of official documents, the database would be a major source of information on the Authority

whose opinion would also be added to the information entered by the Competent Authority of the Interested Party.

- *Consultants and Designers Implementing EIA.* EIA reports identifying potential adverse environmental impacts would be prepared by specialized design and consulting companies that employ authorized experts. Their task would be to identify the present state of all potentially threatened environment components, prepare an adverse impact prognosis and identify relevant mitigation and compensation measures. For this purpose, archives containing data on past uses (together with such information as methodology used, models applied, and necessary measurements) would be particularly useful. Such information could be linked to conclusions drawn from the operation of the facility after completion of the project.

- *Local Administrators.* Usually, a successful EIA would require the cooperation of local authorities assigned to the project; they would be entitled to participate in the decision-making process. Local administrators would be interested in whatever regulations are in force and technical aspects of the EIA as they are the authority that makes the final decision.

- *Non-Governmental Environmental Organizations and the Public.* NGOs normally reflect public opinion. Public participation is a very important element in the EIA process. It should be assumed that the public would need information on a particular project as well as archived data at various stages of the process.

- *Mass Media.* Information — especially inadequate information — may pose significant problems in the decision-making process. Making updated information available to mass media would diminish the risk of informational chaos.

4. Content and Structure of the Database

The database would have several categories of information:

Institutions involved
Names and addresses, full name and position of the person in charge, telephone, fax numbers, electronic mail addresses, URL address of institutions responsible for parts of the Convention procedures, which information would be provided in the database for the following parties:

- *Point of Contact for Notification*: In accordance with Art. 3 of the Espoo Convention, this is a body that should be notified about a proposed project or activity that is likely to cause transboundary environmental impacts

- *Focal Point for Implementation of UNIECE Convention*: This is the body that notifies neighboring countries about a proposed activity or project.

- *Center of Excellence*: An institution acting on behalf of the focal point for implementation of UNIECE Convention in specific matters related to EIA procedure, such as conduct, methodology used, quantification of impacts, etc.

Project records

Each project, current and archived, would have the following information recorded, including project records that accompany the progress of the Convention procedures:

- *General information*: Project title and description, interested party, type of activity, Economic Commission for Europe (ECE) code for the activity type, year of commencement and completion of the procedure.

- *Procedural aspects*: Reference to the information on a proponent and competent body, countries affected by adverse impacts, date of notification and attached comments, date of confirmation and attached comments, date of a public hearing, date and content of the opinion, time for preparation of EIA report, date for EIA report transfer, period for public consultations, period for consultations with interested parties in other countries, date of final decision and its content together with EIA summary report.

- *Comments of affected countries*: In accordance with the Oslo Conference Protocol, an affected party should be allowed to control the way the EIA procedure is performed.

- *Significance of the impact*: Geographical region, special environmental conditions, impact characteristics (size, scope and nature, place, probability, duration, frequency, reversibility), political context.

- *Monitoring the projects*: For archive projects (i.e. the implementation of which had been completed) information will be attached concerning the actual dates of commencement and completion, description of monitored impact (size, scope and nature, place, probability, duration, frequency, reversibility), and final comments.

Legal aspects
A collection of legal regulations concerning EIA and environmental protection issues in international and national context.

Legislation overview
Information written in non-legal language about the international system of environmental law and EIA, the complete text of the Espoo Convention (together with a list of projects and their assigned code numbers), a list of states that signed or ratified the Convention, texts of European Union Directives relating to EIA, and texts of other international conventions on environmental protection.

National EIA legislation
Information written in non-legal language about the system of national environmental law and EIA together with a list of local statutes and texts of local statutory actions concerning EIA.

Bilateral and multilateral EIA agreements
Lists (and possibly complete texts) of bilateral and multilateral EIA agreements which impact on projects within the purview of the Convention.

Additional information
Includes information necessary for successful application and administration of EIA projects:

- *EIA preparation assistance:* EIA training courses, methodologies, research, and literature to assist experts in implementing EIA.

- *Guidelines for EIA Projects:* Brief description of methodologies and sources of additional information which is divided into two categories: general methodology (assessment phases, checklists, scaling techniques, data aggregation, and decision making) and detailed methodologies on such matters as air, water, soil, noise and living nature.

- *EIA training courses in UN/ECE countries:* Technical scope of EIA training courses, organizing agency, venue, dates, how to participate, and language.

- *Research Initiatives:* Qualitative and quantitative methods, cause-and-effect analyses, control and analysis EIA implementation, environmentally friendly alternative solutions, EIA principles on a macroeconomic level, other initiatives such as information on EIA publications and conference proceedings with brief summaries.

5. Database access via WWW and description of the access environment

The best access to the EIATC database will be via the World Wide Web (WWW). This statement is predicated on the assumption that users' means for entering and using the EIATC database will be through WWW once local line transmissions allow efficient use of the Web because the only requirement, apart from an Internet-connected computer, will be appropriate software (e.g. Netscape, Mosaic, Lynx, Explorer).

The database, whose design is completed and which should be operational within two years, has several interesting characteristics. The management of the database is the responsibility of three officials: the technical administrator who is responsible for day-to-day operations and works at the actual site of the database; the country database operator who is authorized to modify and/or add new records from the country he/she represents; and the database content consultant who is authorized to modify data as required.

Among its other features, the database offers users a straightforward process of authorization for using the database, comprehensive on-line help at all stages of use, continuous updating, a system of queries for searching and accessing data, a procedure for modifying or adding data, a means for sending comments to the database administrators, and, as previously indicated, a choicc of working in either English or Russian. Searches may be made by country and within that category, by projects, institutions, legal documents, legislation, research, training courses, and sources of project assistance.

The user who enters the database will be offered several options: general information, query forms, a "what's new" page, a "comments to the administrator" page, and an opportunity to switch languages. There is a built-in method for protection against unauthorized access to any protected data. This involves a process of authorization and passwords. Information supplied by a user will also be subject to a process of verification that requires the user's name, password, and Internet address. Application to use the database must be made through an authorized national institution.

As stated, the design is completed; presently, the computer centers in Geneva and Warsaw are in the final stages of making the database operational by mid-1999.

DISTRIBUTED DATABASE SYSTEMS IN THE MANAGEMENT OF THE ENVIRONMENT AND WATER: ALBANIA

ARBEN MEMO
Institut Fur Anorganische Chemi
Engsbachstrasse 56, App 222
Siegen 57076 GERMANY
Arben.memo@student.uni-siegen.de

Abstract

Albania is the richest country in water resources per capita in the Central and Eastern Europe. However, only one percent per year of those resources are withdrawn for utilization in the major water sectors — drinking, irrigation and drainage, industry, and mining. The management of waters is closely connected to the collection, processing, and distribution of data related to available water resources, their uses, and level of pollution. This study describes the various institutions responsible for information and data on water resources, and their poor quality. Also examined are recent changes in water legislation which have affected the collection, distribution, and access to information on water. The author argues that Albania's involvement in several international projects (e.g. Coastal Monitoring Area Program within the frame of the MED POL Project) requires the application of Quality Assurance and Quality Control (QA/QC) procedures and also offers some recommendations for the establishment of a better data storage and disseminating system and the institutional strengthening of procedures.

Key Words

Albania, data, environment, information, monitoring, networks, pollution, quality, water

1. General Data on Albanian Water Resources and Their Quality

1.1. INTRODUCTION

Albania is a small country covering 28,748 km^2 along the eastern coast of the Ionic and Adriatic Seas. It has a population of 3.2 million inhabitants, 25% of the territory is land, 35% forest, 15% pasture, 4% is covered by lakes, and the remaining two thirds

T. Naff (ed.),
Data Sharing for International Water Resource Management: Eastern Europe, Russia and the CIS, 153–168.

of Albania consists of mountains. Albania has a coastline which is 476 km in length, and along this coastline, covering the western part of the country, lies a small plain with a maximum width of 60 km. Agriculture and forestry together contribute more than 55% to the GDP. The rest is made up of light and heavy industry, food production, oil, mining, the energy sector, and other activities.

Albania's per capita water resources, at a little more than 3000 m^3 (or expressed as specific discharge is 28.1 l/sxkm), are the most abundant in the Central and Eastern Europe (almost equaling Switzerland's). The hydrographic basin of Albania covers a surface of 43,305 km^2 and includes eight large rivers, about 150 small catchments with a surface of more than 50 km^2, and the lakes of Shkodra, Ohrid, and Prespa. There are also a great number of small lakes including the 82 lakes of Dumre, the 12 Lakes of Lura, and the Lakes of Allamani, Gramozi and Klenije. There are also more than 150 springs with a discharge over 120 liters per second (l/s). [1] Until recently, only one percent per year of those resources is withdrawn for utilization in the major water sectors — drinking, irrigation and drainage, industry, and mining. Only two water purification plants exist in the entire country: in Durres (the Pjeshkeza plant on the Erzen River), and Tirana (the Brari on the Tirana River), both built some fifty years ago. Both plants have a treatment capacity of 35 liters/second, while in Tirana the average flow ranges between 1450 and 2100 liters/second and in Durres the flow is about 720 liter/second. Drinking water is disinfected manually with hypochlorite, but a shortage of hypochlorite has contributed to a deterioration of drinking water quality throughout the country.

1.2. THE POLLUTION OF RIVERS

Albania's rivers are subject to both point and non-point sources of pollution. Non-point sources stem chiefly from agricultural activities because pesticides and fertilizers have been used extensively in Albanian agriculture in order to increase productivity in conditions of limited arable land. Besides agricultural activities, another major contribution to water pollution is the waste discharge from different industries (point sources of pollution). The water quality of Kiri, Tirana, Shkumbini, Gjanica, and Dunavec Rivers receive the greatest harm from these discharges because many industrial facilities and several cities are located on or near their banks. (Several characteristics of discharge from some main industrial activities in the country are presented at Appendices 1 and 2.) After 1991, owing to the radical political changes in Albania, many industries interrupted their activities. Because of these circumstances, the deterioration of these rivers has been checked and it is just possible that their general quality has improved; however, this statement must be confirmed by research.

1.3. THE POLLUTION OF COASTLINE WATERS

Both point and non-point sources contribute to the pollution of these waters as well. At the end of 1994, the UNEP, with a view to conducting an inventory of land-based sources of pollution in the four coastal districts of Durres, Lezha, Saranda, and Vlora, distributed questionnaires to Albanian authorities. The questionnaires concerned: (a) liquid domestic discharges; (b) industrial discharges containing selected substances listed in Annexes I and II of the land-based sources protocol within the frame of the MED

POL program; and (c) industrial discharges of petroleum hydrocarbons. For this latter category, sub-questionnaires on discharges from oil-refineries and reception facilities respectively were included. These questionnaires were repeated in early 1995. These particular areas were chosen because of their highest possible impact on coastal waters (including estuaries) through industrial discharges or agricultural activities (The results obtained are tabulated in Appendix 2). The principal findings of the survey are:

- Discharges from municipalities and industry occur off the immediate coastline.
- Protective submarine structures are lacking
- No treatment facilities exist for either sewage or industrial waste water
- No analytical data on the constituents of the waste water are reported
- Mostly untreated industrial water is mixed with sewage thereby aggravating the pollution at the points of discharge
- Untreated wastewaters are discharged mostly into the sea.

1.4. EXISTING SURFACE WATER MONITORING EFFORTS

Since 1948, surface waters (mainly rivers) in Albania have been assessed for hydrometric parameters, since 1966 for hydrochemical parameters, for several other pollutants since 1984, and for estuarine pollution since 1993. There are 100 hydrometric stations in rivers and lakes which study the spatial and time distribution and other characteristics of water resources. There are two categories of stations, class A and class B. In class A stations four kind of observations are made: water level (H), discharge (Q), sedimentation (S), and water temperature (T); at some of the class A stations, water samples for chemical analyses (C) are also taken. In class B stations, generally two kind of observations are made: water level discharges and sedimentation/water temperature.

The hydrochemical analyses to which rivers and lakes are subjected are for pH, Ca, Mg, Na, K, HCO^3, CO^3, Cl^-, SO_4^{2-}, NO_2^-, NO_3^-, SiO_2, P_2O_5, and Fe. They are measured at 45 stations with a frequency of more than 5 samples/year. Samples are generally taken at the hydrometric sampling stations. Between 1984 and 1988 the number of hydrometric stations increased from 10 to 31 making it possible to measure more rivers twice a year for such pollutants as pH, O_2 BOD, COD, NO_2^-, NO_3^-, NH_4, P_2O_5, SiO_2, Zn, Cr, Mn, and Fe. The Tirana and Shkumbini rivers, because of the great amount of effluent discharges they receive from industries, are supposed to be sampled once monthly; however, measurements have been irregular, especially during the 1990s.[1]

An agreement has been made on the National Monitoring Program (NMP) between the Albanian Government as the contracting party to the Convention for the Protection of the Mediterranean Sea against Pollution and its related Protocols, and UNEP as the Secretariat to the Convention. Since 1992, within the frame of the Pollution Monitoring and Research Programme (MED POL), Albania has started a monitoring program, for the first time, for estuarine, bathing areas, and pollution sources (industrial and urban discharges and agricultural activities). The Albanian MED POL National Monitoring Program includes 73 sampling stations. Twenty-four are located in the main estuarine areas, nine cover main sources of pollution, and the remaining forty are located at bathing areas. Since the project began in 1993, only one set of results — those for 1994 — have been produced. In May 1996, the UNEP published the following results in its *Report on Results of Coastal Areas Management Program (CAMP) for Albania*:

TABLE 1. Average values for some chemical pollutants in mg/l. UNEP, 1996

	0_2	NBO	LST	NO_2	NH_4	NO_3	PO_4	P Total
Lezha	6.95	0.85	8.4	0.002	None	None	None	None
Durres	7.05	0.39	4.2	None	None	None	None	None
Lushnja	7.30	1.05	6.4	0.001	None	0.002	None	None
Fier	8.32	0.37	10.0	0.005	None	0.060	None	None
Vlora	8.02	0.50	6.0	0.004	None	0.006	None	None
Saranda	7.04	1.48	1.0	0.001	None	0.092	None	None

TABLE 2. Average values of the heavy metals of the marine sediments in mg/kg (dry weight). UNEP, 1996

	Hg	Cd	Pb	Cu	Cr	Zn	Ni	Mn	Fe
Lezha	.070	.136	11.4	255.1	401.1	135	310.7	975	61.2
Durres	.129	.445	29.3	100.7	263.0	65.9	189.9	614	24.2
Lushnja	.079	.251	14.9	63.1	283.0	95.4	376.0	470	57.8
Fier	.053	.186	14.5	31.6	298.0	122	303.0	499	44.5
Vlora	.773	—	—	33.6	314.0	47.6	—	512	35.1
Saranda	.041	.272	—	47.1	215.0	108	216.0	410	33.3

UNEP specialists drew some rough conclusions from these results. Based on the data in Table 1, they state that the estuaries are not influenced by the discharges of various communal sources. From the data in Table 2, they concluded that the levels of heavy metals in the majority of samples taken are comparable to values in other less polluted areas of the Central Adriatic [3].

The UNEP conclusions provide only a rough explanation of the water quality situation at these particular sites for the following reasons:

- Water quality trends could not be detected through NMP because it is designed only for background monitoring. This design was determined by the lack of general data on the quality of these waters
- Of the three planned data sets, only one has been produced and even in that one the results were incomplete. In these conditions the water quality situation and quality trends are impossible to determine
- A lack of institutional co-ordination and legislative requirements have hindered the full implementation of the monitoring program
- Like other monitoring efforts undertaken in Albania, this program lacks clearly defined monitoring objectives
- A clear plan for effectively communicating results to those who have need of the data is absent from the program.

2. Present Water Uses in Albania

The chief water supply uses in Albania are for drinking, mining and industry, irrigation and drainage, hydropower, and other non-consumptive uses such as tourism, recreation, aquatic life, fisheries, and flood mitigation.

2.1. DRINKING WATER SUPPLY

Drinking water supply is a national priority because this service involves the well-being of the entire population and requires improvements. Drinking water supply is responsibility of the Ministry of Public Works, Territory Adjustment, and Tourism (MPWTAT). In the main urban areas 80% of the population is connected to the network of water distribution, but this service can be used only a few hours a day. The pipes, which are made of steel and cast iron, were installed some 20 or 30 years ago. The lack of preventive and routine maintenance has resulted in their becoming severely corroded. In general the sources of drinking water are of good quality but the tap water may be contaminated by contact with chemical and bacteriological agents in the soil. There is a lack of information concerning rural water supply. The 1995 UN Human Development Report on Albania states that only 5% of the rural population has access to in-house water pipe connections. Also, according to the same report 98% of the rural population rely on water near their homes which is piped or extracted from shallow wells where water is of good quality.

2.2. MINING AND INDUSTRIAL WATER USES AND DISCHARGES

Mining and industry have been very important users of Albania's water resources during the last 30 years. However, since 1991, a period of political and economic transition, mining and heavy industrial activities have decreased. Although the quantity of untreated water discharged by these industries into the environment has diminished, the effects of their previous dumping are still evident, especially in the large urban areas of Tirana, Fier, Korca, and Durres. The main point sources of pollution for surface and ground water are the extraction of chromium near Dnni and Mati, copper extraction near Dnni, Mati, and Seman, oil extraction near the Vjosa River, wherever the chemical industry produces such products as fertilizer, paints, oil products, and the metallurgic industry, paper industry, textile and leather industries etc. (Figures for the discharges of Albanian industries are given in Appendix 1 and Tables 3 & 4 below.) The information concerning the quality and quantity of water used by industries and mining activities is so poor as to be almost lacking altogether.

TABLE 3. Types of industrial discharge, method of discharge and total amounts for coastal zone surveyed.

	Industrial wastewater		Cooling wastewater		Domestic wastewater	
Discharge Method	Untreated mcm/yr	Treated	Untreated	Treated	Untreated	Treated
Municipal system	13.05	0	N/A	0	8.88	0
River	0.9	0	N/A	0	0	0
Onshore	1.2	0	N/A	0	0	0
Other	0	0	N/A	0	0	0
Total	15.15	0	N/A	0	8.8	0

TABLE 4. Types of industrial discharge, method of discharge and discharge amounts per district surveyed.

District	Industries	Total wastewaterin m^3/day	Domestic wastewater	Industrial wastewater		
				municipal system	river	on-shore
Durres	Brewery Distillery Pharmaceutical Naval Stores Pesticides Rubber	40000	17000	23000	0	0
Lezha	Pulp/Paper	7800	1300	3500	3000	0
Saranda	Bakery	2300	1300	1000	0	0
Vlora	Tannery Acid Plastics/Resins Cement	30000	10000	16000	0	4000
Total		80100	29600	43500	3000	4000

2.3. IRRIGATION AND DRAINAGE

As Albania is an agricultural country (about 55% of GDP comes from agricultural activities), great attention has been paid to irrigation and drainage works. The rate of investments made in the agricultural sector between 1950 and 1975 significantly increased irrigation capacity. Albanian specialists have estimated that owing to the social upheavals following the 1991 fall of communism in Albania, 14,400 ha of the country's irrigation systems and 15,300 ha of the drainage systems were severely damaged. Moreover, according to rough estimates, the present water diversions satisfy only 25% of the irrigation needs. The improvement of the irrigation and drainage systems is imperative if there is to be a food supply adequate to meet the demands of both the population and the food processing and export industries.

2.4. HYDROPOWER

Large resources in surface waters and high river bed slopes have favored Albania with a large potential for hydropower. The current installed capacity is 1446 MW, the annual production is capacity is 5220 Gkh, and the water storage capacity is approximately 4x109 m^3. There are a total of 12 hydropower plants located on several rivers: three on Dini River, two each on the Mati and Bistrica Rivers, one each on Erzeni, Bogova, Osumi, and Shushica Rivers, and one under construction on the Devolli River. The Drini cascade in northern Albania contributes 350 MW, or 93% of total energy production. The dams at the hydropower plants are designed mainly for hydropower generation. The first multipurpose dam, currently under construction on the Devolli River will enable irrigation in the Plain of Semani. Other non-consumptive uses of water in Albania are aquatic life and fisheries, tourism, recreation, navigation, and flood mitigation. Some of these activities cause limitations on both consumptive uses and upstream discharges because of the water quality they require.

3. Institutional Aspects of the Current Water Data Information System

3.1. ENVIRONMENTAL ADMINISTRATION

Environmental information systems in Albania, including the water data information system, have been in a state of evolution since 1991. In that year the Committee of Environmental Protection (CEP), which is based in the Ministry of Public Health but under the jurisdiction of the Council of the Ministers, was created. The CEP is the central body for all environmental issues. One of its functions is to organize the monitoring of environmental (and water) pollution and, in cooperation with National Water Council to define what actions are to be taken for the protection of water resources. Another responsibility of CEP is to approve the legal standards and criteria for hazardous discharges, in both solid and liquid forms, into the country's waters. There are three Directorates in the CEP, one for water and air quality, another for natural resources protection and the management of hazardous materials, and a third for a national information system and impact assessments.

In 1996, the Albanian Parliament passed legislation that made the National Water Council (NWC) the highest decision-making body responsible for the administration and management of water resources in Albania. The NWC is appointed by the Council of Ministers and the Prime Minister is the NWC chairman. The National Water Council's chief responsibilities are:

- the development and management of a national water strategy
- the preparation of water resources plans for drainage basins
- issuing of regulations for the implementation of the water law and national water strategy
- research concerning the efficient use of water resources in Albania
- organization and application of an administration for the implementation of water law.

The approval of the water law has lead to some uncertainties about future responsibility for some related activities; for example, the publication of regular reports on water quality. In addition to CEP, there are several ministries that deal with various issues related to water management, e.g., the Ministry of Public Works, the Ministry of Territorial Adjustments and Tourism, the Ministry of Food and Agriculture, the Ministry of Transport, the Ministry of Industry and Energy, and the Ministry of Public Health. Currently, owing to a shift towards a market economy, the CEP and other governmental agencies related to water issues are losing some of their control and management functions. The consequence for data collection and management is that instead of having data controlled exclusively by the central authority, data must now be requested from private enterprise and individuals.

The question of financing the collection and distribution of water information is partly related to the implementation of the Polluter-Pays-Principle (PPP), by which the costs of water pollution monitoring is to be borne by the polluter. In situations where the polluters cannot be identified (e.g. non-point sources of pollution), water data gathering is to be publicly financed. However, the regulations governing PPP are not strongly implemented in Albania. Circumstances do not favor the application of a modern system of fines and taxes against the polluters.

3.2. THE ROLE OF INSTITUTES

There are in Albania several scientific institutions that cooperate with CEP and the ministries on matters of water resources management. As part of the privatization of the economy, many such institutes have become administratively and financially separate, semi-independent organizations. The principal water-related institutes in Albania are:

- The Institute of Hydrometeorology of the Academy of Science which conducts water research and monitors water quality
- The Institute of Public Health which investigates and controls at the national and local levels the quality of water for domestic supply and establishes regulations and guidelines for maximum levels of toxic substances which may be tolerated in domestic water use
- The Faculty of Natural Sciences which conducts research on water treatment and is also involved in several projects dealing with water quality assessment in some areas of Albania
- The Research Institute of Chemical Technology inventories urban and industrial water discharges throughout the country.

The most dynamic and well-equipped institutes are attempting to establish themselves as the focal points of international information networks. Until the CEP began functioning and after the legislation of the Water Law, there was a tendency for projects to be undertaken based on their attractiveness to external funders rather than on real national needs. This tendency has been minimized by the influence of CEP and the new water law.

4. Technical Aspects of the Water Data and Information System

4.1. THE SELECTION OF ANALYTICAL METHODS AND QUALITY ASSURANCE/QUALITY CONTROL PROCEDURES

In Albania, as elsewhere, the availability of analytical equipment has influenced the choice of analytical methods. Albanian specialists have always been eager to use the most up-to-date analytical techniques in order to ensure high reliability of data. The Institute of Hydrometerology of the University of Tirana and the Institute of Chemical Technology are among the institutions which now appear to be well-equipped due to governmental and international funding from such UN agencies as UNEP and UNESCO. This circumstance has made it possible at these two institutes to select optimal analytical methods based on their sensitivity, specificity, response-time, ease of operation, ease of calibration, cost and reliability, and accuracy. Until recently, there have been virtually no general quality control or quality assurance procedures implemented. Under the requirements of MED POL projects, the laboratories involved are assessed for their accuracy and precision only through intercalibration exercises.

4.2. DATA STORAGE, REPORTING, AND DISSEMINATION ISSUES

In the past, collected water data was stored only in hardcopy form. A computerized system of storage is under development. Because of a lack of funding for computerization, only in specific cases are data put into computerized files, but systems for processing them are not yet installed. Current methods of storing water quality data make access to them difficult for the scientific community and general public alike, and hinder the flow of data to decision-makers. These "hidden data" have led to duplication of efforts and studies on water quality issues by various institutions or interested parties.

The reporting of water quality data is the duty of the Committee for Environmental Protection which is required to submit to the parliament each January water quality findings as part of its annual report. Until recently it has been quite difficult for the CEP to collect all the data on the quality of the environment generated by the different institutions, even though it is required by Article 33 of the Albanian Environmental Law:

> *Information on the environmental situation shall be received and stored by the Committee of Environmental Protection and its regional agencies, by other ministries and central institutions, and councils of communes, municipalities or districts according to the relevant territorial units. Official and non-official persons shall be obliged to forward information on the environmental situation within two weeks from the date a request is received. The information must be forwarded to the competent authorities in accordance with procedures defined by the Minister of Health and Environmental Protection.*

In these circumstances, it has not been possible to coordinate monitoring efforts. This has created difficulties for the CEP in fulfilling its reporting obligations. To correct this situation, the Albanian government issued an edict (No. 541) on September 25, 1995. Article 6 of this edict mandated that several responsible ministries, institutions, officials, and others entities must submit monitored data each January to the CEP.

According to the Article 34 of the Environmental Law, the above designated institutions are also obliged to make the information they own accessible to the general public, but until the present, they have failed to comply with this rule. The normal format for reporting are standard yearbooks or reports. The most regular scientific publications have been issued by the Institute of Hydrometeorology including the biannual *Journal of Hydrology*. Other forms of publication — booklets, newsletters, specialized reports on various specific issues, environmental campaigns on TV and other media — are quite rare, often produced only when an international organization funds or requires them, e.g., the *Albanian Environmental Strategy Report* (World Bank, 1993), *Reports on the Results of Coastal Areas Management Program (CAMP) for*

Albania (UNEP, 1996). The frequency and dissemination of these reports depends on the requirements of the respective projects.

4.3. THE RELEVANCE OF ENVIRONMENTAL AND WATER INFORMATION FOR DECISION MAKING

There are two main conditions necessary for making environmental and water information relevant to decision-makers. First, the information should be adapted to the type of decision to be taken at various levels, e.g., the project level, for environmental auditing, budget and planning, or for reporting on the general state of the environment. Second, the information should match the type of decision-maker, that is, whether the decision-maker is someone in the private sector or in the public sector, e.g., the CEP.

There is general dissatisfaction among the chief governmental institutions with the quality and relevance of the environmental information available from the CEP. Within individual Albanian ministries and other government entities, established channels for the flow of information from providers to users exists. These institutions have their own sources for most types of environmental information which they use for their own internal policy decisions. For example, the Ministry of Mining, Industry, and Energy has created the following internal structures through which necessary information is transmitted throughout its own units

DIAGRAM 1. Environmental Network Established at the Ministry of Mining, Industry and Energy Resources

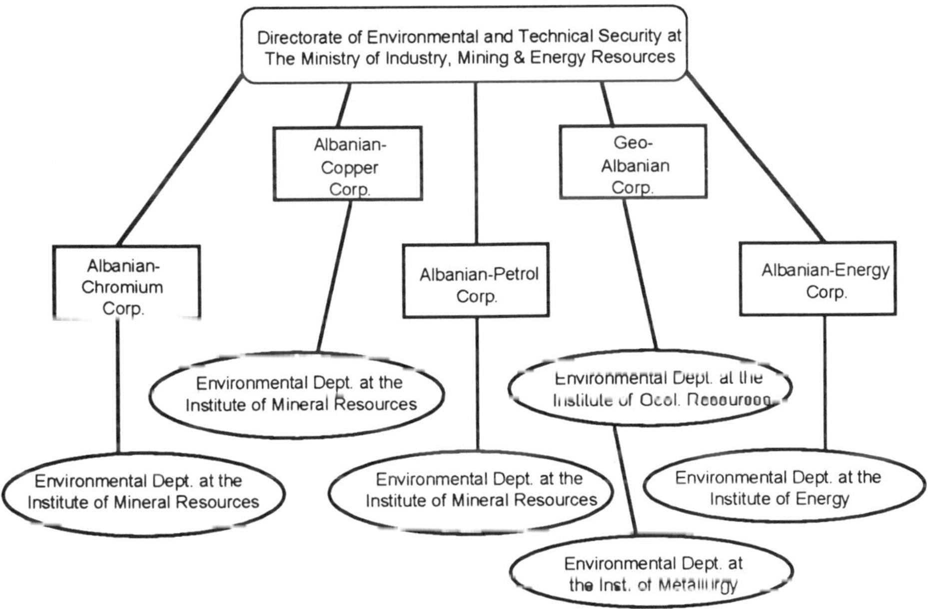

Consequently, in the absence of a rational system for an inter-governmental exchange of information, difficulties arise when information is needed by other ministries and government agencies. The work of decision-makers is, thus, seriously hampered. There are no quality or relevancy checks on information that does pass from one agency to another. Nor is there a central coordinating unit such as a National Statistical Office that could collect, process, and manage information for all government and private sector units ensuring that the best and most relevant information reaches the decision-makers who need it in a timely fashion.

5. Conclusions

At present, water data and information in Albania are very difficult to collect and share because they are scattered among many ministries, agencies, and institutes. In addition, much of the data is still treated as being secret. It is virtually impossible for the general public to access this information.

In general, data related to water issues is considered by individual specialists and scientific institutions as private preserves and their delivery is conditional upon payments of money. Even the flow of information between governmental agencies is often subject to the same conditions.

An improved water database system would allow better management of the nation's water resources, lead to more rational and efficient policy choices such as how to improve the water supply, the treatment of urban and industrial wastewater and where best to discharge it.

Improvement of current data issues will require:

- establishment of computerized data storage systems
- entry of hard-copy data
- standardization of data nomenclature and definitions
- employment of a meta-data system that describes the content, quality, nature, and other characteristics of data
- combination of centralized and distributed database networks
- extensive use of the Internet and World Wide Web for data sharing.

Although the advantages of computerized data storage and retrieval systems are obvious, it is very important for developing countries to conduct a careful cost-benefit analysis before implementing such systems. (Diagram 2 shows factors necessary for making the best choices.)

DIAGRAM 2. Evaluation of costs and benefits for choosing a computerized data storage and retrieval system

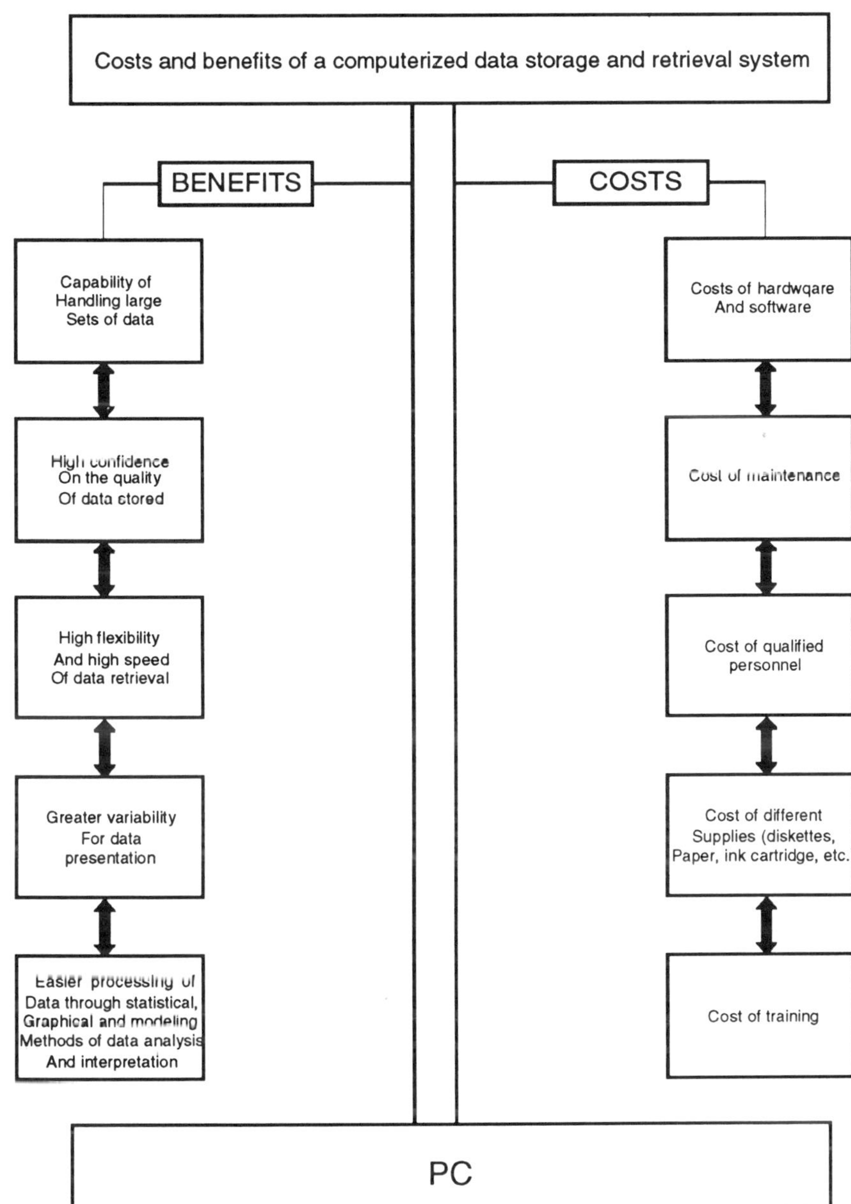

The following general recommendations are offered for the improvement of reporting and the dissemination of data and information:

- the format and frequency of dissemination should be determined during the design phase of water management systems
- reports should be aimed at particular constituencies
- reports should be formatted in conformity with the requirements of public or private sector decision-makers or of the general public
- graphs, diagrams, and tables should be an integral part of the reports, especially when they are addressed to the general public.

Notes

1. BCEOM (1996) *National Water Strategy for Albania-Inception Report,* BCEOM, 35.

2. UNDP (1996) *Human Development Report - Albania 1995,* Tirana, United Nation Development Program, 54.

3. UNEP (1996) *Report on the Results of Coastal Areas Management Program (CAMP) in Albania,* Athens, UNEP, 45.

4. UNESCO/WHO/UNEP, (1992) *Water Availability Assessment. A guide to the use of Biota. Sediments and Water in Environmental Monitoring,* Cambridge, U.K., University Press, 589.

APPENDICES

Appendix 1.

TABLE 1. Some chemical pollutants discharged from main industrial sources in Albania from BCEOM (1996) National Water Strategy for Albania – Inception Report, BCEOM, page 35.

Main Con-stituents	Some major industrial activities					
	Pesticide factory in Durres	Chemical factory in Tirana	Leather processing in Korca	Textile compound in Tirana	Oil processing compound in Ballsh	Effluent standards (mg/l)
SO_4^{2-}		15 g/l				500
F^-						251
As						0.5
Cl^-	30g/l					1500
TSS			1200 - 2800 mg/l	78 – 515 mg/l		100
Total chromium			14 – 16.2 mg/l			3
Cr^{6+}	23.8 g/l					0.5
S^-	15 g/l					5
Phenol						40
Form-aldehyde		7 mg/l				2
Benzene		2 mg/l				
BOD_5			1000 – 1600 mg O_2/l	19 – 315 mg O_2/l		15
COD			4900 – 7200 mg O_2/l	200 – 1000 mg O_2/l		50
Discharge	Adriatic Sea	Tirana River	Dunavec River	Tirana River	Gjanica River	
Quantity		2 m^3/hr	500 m^3/day	3500 m^3/day	3,000,000 t/yr	

Appendix 2.

TABLE 1. Population, water consumption and wastewater quantity

District	Population		Water Consumption in m^3/month	Wastewater		
	Normal	Seasonal increase		Production $10^6 m^3$/yr	Discharge into sea or river $10^6 m^3$/yr	Percentage discharged
Durres	75,000	35,000	500,000	4.8	4.56	95.0
Lezha	12,000	2,500	40,000	0.41	0.37	90.24
Saranda	12,000	3,500	40,000	0.41	0.39	95.12
Vlora	75,000	25,000	300,000	2.9	2.6	89.66
Total	174,000	66,000	880,000	8.52	7.92	92.96

TABLE 2. Municipal wastewater deposition

District	Total urban wastewater $10^6 m^3$/yr	Estimated annual discharge in million m3/yr				Wastewater re-use
		Into sea (municipal sewage system)	On land	In subsoil	Other	
Durres	4.80	4.56	0	0.24	0	0
Lezha	0.41	0.37	0	0.04	0	0
Saranda	0.41	0.39	0	0.02	0	0
Vlora	2.90	2.60	0	0.30	0	0
Total	8.52	7.92	0	0.60	0	0

WATER QUALITY DATA COLLECTION AND SHARING BETWEEN HUNGARY AND NEIGHBORING COUNTRIES

FERENC LASZLO
Water Resources Research Centre (VITUKI)
H-1453 Budapest, P.O.B. 27, HUNGARY

Abstract

Hungary is a typical downstream country within the Danube basin. The major part of its surface water resources originate outside its borders. Hungary's national boundaries are crossed by 90 water courses. Consequently, Hungary has a great, on-going interest in water quality monitoring of the transboundary rivers.

Bilateral agreements with the neighboring countries ensure the legal basis for regular joint water quality investigations, water quality evaluation of transboundary rivers, and water quality data sharing across national boundaries.

The recently implemented *Transnational Monitoring Network* (TNMN) and *Danube Accident Emergency Warning System* (DAEWS) which were developed within the Danube Environmental Program provide coordinated, up-to-date water quality monitoring and warning methods for the Danube countries.

Key Words

DAEWS, Danube River, data, environment, Hungary, pollution, quality, TNMN

1. Introduction

Hungary, the greatest part of whose 93,000 km^2 of territory is covered with plains, is situated in the deepest part of the Carpathian basin. The whole of the country belongs to the Danube River system. Because of its geography, the waters rushing down from the surrounding higher elevations of the Carpathian basin converge in Hungary. All of the major streams that flow through Hungary originate beyond the country's borders, including 94 percent of surface water sources. One third of the national boundaries themselves are constituted by streams and rives over a length of 600 km. These borders are traversed by 90 water courses of varying size.

T. Naff (ed.),
Data Sharing for International Water Resource Management: Eastern Europe, Russia and the CIS, 169–177.

2. Surface Water Quality Monitoring in Hungary

National water quality observations began several decades ago. In practice, monitoring, which is regulated by statute (the National Standard MSZ 12749), occurs nominally at biweekly intervals, but in the key hydrological cross-sections samples are gathered each week. The total number of sampling sites is 150. The samples are analyzed for the main cations, anions, nutrients, trace elements, organic micropollutants, radiological components, bacteriological and hydrobiological parameters.

The monitoring network, which is operated by the environmental and public health authorities, is required to perform the following functions:

- provide data needed for the general assessment of the quality of surface waters,
- to build up a database suited to monitoring any changes in the quality of surface waters,
- detect and identify pollution impacts originating from natural sources, production, and consumer activities in Hungary and abroad,
- produce the data series needed for international water quality assessments and negotiations on boundary streams, and
- produce data series for research and planning purposes.

The standard sampling stations for the monitoring of surface water quality are situated:

- in the boundary cross-sections of streams entering or leaving the country,
- upstream and downstream of discharges affecting water quality in streams, and
- in cross sections of special importance (e.g. water intakes).

Other important considerations in selecting the sampling sites were coincidence with a station of the hydrological observation network, or if this was not feasible, then where there was a possibility of establishing valid correlation with the nearest gauging station, and, accessibility. The sampling points are, by and large, located in the stream centerline or in the line of the main current.

The factors involved in deciding upon the frequency of sampling have been the data demand of statistical processing, the variability of water quality in time, the importance of the sampling site, and the capacity of the analytical laboratory facilities involved.

The samples are analyzed in the laboratories of the 12 District Environmental Inspectorates for most of the chemical and hydrobiological components that might be or are present in the water. However, these district laboratories do not have the technical proficiency to analyze all the monitored micropollutants. Therefore, specific organic micropollutants are analyzed in appropriately instrumented central laboratories. Microbiological parameters are measured by the Public Health Laboratories of the nation's 20 counties. Since these laboratory analyses are part of the water quality monitoring process, the analytical methods employed are also regulated by the MSZ 12749 Standard. Special attention is paid to the quality assurance of the analytical work of the laboratories by running an intercalibration program.

The measured data are sent from the laboratories on diskettes at monthly intervals to the Institute for Environmental Management (KGI). The collected data are then checked and stored in the national database. A special software, called VM, was developed to process the database. VM can provide statistical evaluations of the data in accordance with the classification system MSZ 12749.

3. Water Quality Monitoring in Transboundary Waters

3.1. BILATERAL AGREEMENTS

Owing to Hungary's hydrogeographical situation, cooperation between Hungary and its immediate neighbors with whom it shares most of its waters, is of vital importance to the collection and sharing of water quality data among all the Danube's riparians. Cooperation in transboundary water management is regulated chiefly by bilateral agreements. Joint Commissions on Transboundary Waters, established for the purpose, are responsible for maintaining and carrying out the collaboration.

Joint efforts undertaken regularly, such as water quality investigations, quality evaluations, and cooperative measures against accidental pollution began in the 1960s and 1970s. At present, regular water quality investigations are carried out jointly with neighboring countries for the water quality parameters listed in Table 1. Some technical details of the bilateral transboundary monitoring practice are summarized in Table 2.

TABLE 1. Components monitored in transboundary waters.

Component	Transboundary Waters				
	Austria - Hungary	Slovakia - Hungary	Ukraine - Hungary	Romania - Hungary	Former Yugoslavia - Hungary*
Dissolved oxygen	+	+	+	+	+
DO saturation	+	+	+		+
BOD2	+			+	
BOD5	+	+	+	+	+
CODp	+	+	+		+
DOCd		+			
TOC	+				
DOC	+				

TABLE 1. (continued)

Component	Transboundary Waters				
	Austria - Hungary	Slovakia - Hungary	Ukraine - Hungary	Romania - Hungary	Former Yugoslavia - Hungary*
Ammonium	+	+	+	+	+
Nitrite	+	+		+	
Nitrate	+	+	+	+	+
Organic Nitrogen		+			
Total N		+			+
Ortho-phosphate	+	+		+	+
Total P	+	+			+
Chlorophyll-a	+	+		+	
Arsenic	+				+
Zinc	+	+			+
Mercury	+	+			+
Cadmium	+	+			+
Chromium	+	+			+
Nickel	+				+
Lead	+	+			+
Copper	+	+			+
Phenols	+	+	+	+	+
MBAS-s	+	+	+	+	+
oil		+	+	+	
Atrazine	+				
AOX	+				
PAH-s	+				
PCB-s	+				
Gross B activity			+		
Ph	+	+	+	+	+
Conductivity	+	+			
Iron	+	+	+	+	+
Manganese	+	+	+	+	
Water temp.	+	+	+	+	
Air temp.	+				
SS.	+	+	+		
TDS		+	+	+	+
m-alkalinity	+				
P-alkalinity	+				
Total hardness	+	+	+		+
Carbonate hardness	+				
Calcium	+		+		
Magnesium	+		+		
potassium	+		+		
Sodium	+		+		
Chloride	+	+	+		+

TABLE 1. (continued)

Component	Transboundary Waters				
	Austria - Hungary	Slovakia - Hungary	Ukraine - Hungary	Romania - Hungary	Former Yugoslavia - Hungary*
Sulphate	+	+	+		+
Hydrogen-Carbonate	+				
Color	+				
Turbidity	+				
Odor	+				
Coliform	+	+	+		+
Faecal coliform	+				
Faecal streptococcus	+				
Salmonella	+				
Plate count (22oC)	+				
Saprobity Index	+	+	+		+
Algae count					
Water level	+				
Water discharge	+			+	

TABLE 2. Details on water quality monitoring of transboundary water.

Countries	Austria-Hungary	Slovakia-Hungary	Ukraine-Hungary	Romania-Hungary
Number of sampling sites	11 + 8** = 19	10 +1 x 3** = 13	2	7
Sampling procedure	In odd months: Austrian sampling In even months: Hungarian sampling	Joint sampling	Joint: 4 Ukranian: 4 Hungarian: 4	Romanian: 12 Hungarian: 12
Sampling frequency per year	12	12	12	24
Frequency of analysis (per year)	Usually: 12 Special components: 6	Usually: 12 Special components: 6	12	24
Laboratory site of analysis	Odd months: Austria Even months: Hungary	Slovakia: 12 Hungary: 12	Ukraine: 8 Hungary: 8	Romania: 12 Hungary: 12
Number of yearly data agreement meeting	1	2	1	Results accepted without data agreement meeting
Data exchange	Direct exchange of hard copies	Direct exchange of hard copies	Direct exchange of hard copies	Data exchange by fax
Evaluation method of the measured data	Statistical parameters biyearly	Statistical parameters buyearly	Statistical parameters in every year	Statistical parameters in every year
Trend analysis	Once in 10 years	Once in 5 years	-	Once in 5 years
Spreadsheet software	EXCEL 5.0	EXCEL 5.0	-	EXCEL 5.0
Water quality criteria	National standards	CMEA system (six classes, 1984)	CMEA system (three classes, 1963)	-
Frequency of interlaboratory calibration	One per 12 sampling	-	One per 12 sampling	One per 12 sampling

TABLE 2. (continued)

Countries	Yugoslavia-Hungary*	Croatia-Hungary*	Slovenia-Hungary*
Number of sampling sites	3	3	1
Sampling procedure	Joint: 4 Yugoslavian: 4 Hungarian: 4	Joint: 4 Croatian: 4 Hungarian: 4	Joint: 4 Slovenian: 4 Hungarian: 4
Sampling frequency per year	12	12	12
Frequency of analysis (per year)	12	Usually: 12 Special components: 4	Usually: 12 Special components: 4
Laboratory site of analysis	Yugoslavia: 8 Hungary: 8	Croatia: 8 Hungary: 8	Slovenia: 8 Hungary: 8
Number of yearly data agreement meeting	1	1	1
Data exchange	Direct exchange of hard copies	Direct exchange of hard copies	Direct exchange of hard copies
Evaluation method of the measured data	Statistical parameters in every year	-	-
Trend analysis	-	-	-
Spreadsheet software	-	-	-
Water quality criteria	CMEA system (three classes, 1963)	CMEA system (three classes, 1963)	CMEA system (three classes, 1963)
Frequency of interlaboratory calibration	-	-	-

* Situation in 1991
** Lake Ferto: longitudinal and cross-sectional sampling sites
*** River Danube at Szob: cross-sectional sampling sites

3.2. THE TRANSNATIONAL MONITORING NETWORK (TNMN) OF DANUBE COUNTRIES

In addition to the bilateral activities, considerable international water quality monitoring is carried out within the Danube Environmental Program. The Danube states jointly operate TNMN which includes 56 sampling sites along the Danube River and in its main tributaries. Water samples are taken each month and analyzed for physical, chemical, radiological, bacteriological and hydrobiological parameters. The analysis is extended to the suspended and bottom sediment as well.

For its part, Hungary operates nine sampling sites in TNMN at the following river sections:

- Danube, GyoQamoly (boundary section)
- Danube, Komarom (boundary section)
- Danube, Szob (boundary section)
- Danube, Dunafoldvar
- Danube, Hercegszanto (boundary section)

- Sio, Sioagard
- Drava, Dravaszabolcs (boundary section)
- Sajo, Sajopuspoki (boundary section)
- Tisza, Tiszasziget (boundary section).

As indicated, seven of the sites are boundary sections. The measured data are collected in standardized data file format and stored in the provisional TNMN information center in Bulgaria. The processed data and their evaluation will be published in yearbooks, starting with the year 1996.

3.3. THE DANUBE ACCIDENT EMERGENCY WARNING SYSTEM (DAEWS)

The Danube riparian states agreed to develop DAEWS as part of the Danube Environmental Programme. The general objectives of DAEWS are to increase public safety by protecting drinking water resources and other water uses if accidents occur in the Danube River or in its tributaries, and to protect the environment against the effects of such accidental pollution.

The specific objectives of DAEWS are:

- prompt joint action on emergencies that may take place in the rivers and tributaries of the Danube River basin and timely exchange information for the purpose

- prompt joint action on unpredictable changes in the water levels that may occur in the rivers and tributaries of the Danube River basin and timely exchange of information for the purpose

- prompt reception, processing, and transmission of information on sudden, accidental incidents of pollution caused by dangerous substances which have entered surface waters and which would cause significant transboundary harm.

When sudden pollution occurs, the system must have the capability of promptly warning the services responsible for handling such emergencies so that they can take quick action to contain the danger, ascertain the cause of the problem, rectify the damage, take steps to avoid the consequences, and find those responsible. Even when minor accidents are involved that might alarm the public, the system must be able to communicate quickly information which the public needs to know in order to evaluate the problem.

In the DAEWS information routing scheme warning messages are sent from upstream to downstream. Hungary's responsibility is to warn the following countries in the event of accidental pollution in the transboundary rivers listed below (Hungary, of course, would be given warning by its upstream neighbors):

Country	Transboundary River
Slovakia	Danube, Ipoly/Ipel
Romania	Danube
Croatia	Danube, Mura, Drava

4. Conclusions

Hungary, as a downstream riparian, is highly interested in the cooperative monitoring of water quality of its transboundary rivers. The existing bilateral agreements constitute a good framework for water quality data sharing with neighboring countries. The recently instituted TNMN and DAEWS systems provide coordinated water quality monitoring and warning networks for all the Danube River countries.

DEVELOPMENT OF THE HYDROLOGICAL SUBSYSTEM OF THE WATER MANAGEMENT INFORMATION SYSTEM IN HUNGARY

ANDRAS TARNOY
Unibox
1026 Budapest
Hidàsz utca 15
HUNGARY

Abstract

The system for processing hydrological and hydrogeological data in Hungary, which was developed in 1989-90, can no longer be revised or reorganized. New communication technology (based on Lotus Notes groupware) which is being utilized, together with a new generation of relational DBMS (MS Access, MS SQL Server) offer an opportunity for planning and developing a new solution to the country's water data management needs. The new approach, which falls within a newly conceptualized information system of water management, involves a highly automated, standardized and shared hydrological information subsystem (HISS).

The process of implementing the new plan required several stages: conceptual planning in 1995, project approval in 1996, followed by base studies — definition of the study, the assessment of existing information applications, and the Conceptual System Plan — then the development of a pilot model of the automated Module for the reception of daily hydrological data and their distribution among the District and Central Water Authorities, followed by the development and installation of appropriate Geographical Information Software, and, finally, the launching of the new Hydrogeological Information Subsystem (or Hydrological Information Subsystem) — HISS.

This study describes the conceptual and topological structures of HISS, summarizes its data and functional requirements, and its GIS user interface.

Key Words

data, database, DWA, GIS, HISS, Hungary, information, NWA, VITUKI, water management

1. Water Management Administraion

State water authorities are legally responsible for the governance of water uses which involves many functions: managing water works, sustainable development, assessment

T. Naff (ed.),
Data Sharing for International Water Resource Management: Eastern Europe, Russia and the CIS, 179–185.

and management of national water resources including the operation of hydrological monitoring, measurements, and data management, the maintenance of flood controls, protection against water quality accidents, poor drainage, and state owned water infrastructure facilities and projects, etc.

The national authorities cardinal responsibility is to ensure good water quality. However, the collection of ground water quality data is still collected by various water management institutions such as the environmental authorities, the state meteorological, geological, and public health services, and regional and local communities. By and large, they cooperate in the execution of their respective duties.

The government body with the greatest control over national water affairs is the Ministry of Transport, Telecommunication, and Water Management (MTTWM). It should be noted that the Ministry's responsibility for water management does not extend to quality control. That is the domain of the Ministry for Environmental and Regional Policy. Communal waterworks institutionally fall within the purview of regional and local authorities, but the professional control of their activities — e.g., assignment of specialists — remains with the MTTWM.

The National Water Authority, which is an agency subordinate to the MTTWM, is the principal administrative agency responsible for water management, and at the next level are the 12 District Water Authorities (DWAs).

In addition to the central and district authorities, a key role in water data processing and information services is played by the Scientific Research Institute for Water Management (VITUKI). The hydrological data monitoring network is decentralized, with control and inspection of the monitoring network falling under district water authorities.

2. Information Requirements

An analysis of the data and information requirements of those agencies related to the water sector demonstrated that a fairly wide range of hydrological information is needed by practically every user group. A similar inventory of data made clear the necessity for identifying the key objectives of all user groups. These considerations induced information planners to cancel a plan for renewing an established hydrological information application package and instead to create a new hydrological information subsystem, with standardized definitions and terminology and root database subsystem as part of a future more complex water management information system.

3. Data Activities and Requirements

Data and Activities	Information Required
Hydrometry	Daily time series; surface data; monitoring network data; user inventory; riverbed geographical data
Hydrogeology	Water use & users; aquifer geographical, geological data; local withdrawals; waste water disposal; industrial water supply; sewage systems; state investments
AgricultureWater	Delivery & consumption; hydrological & agrometeorological data; irrigation system & irrigated crops inventories; water uses
Lakes and Rivers	Hydrological data; river & lake bed geography; inventory of structures; time series flood/drainage; flood/drainage events historical data; pollution sources & protection structures inventory; daily protection measures
Finance/Maintenance	Hydraulic structures inventory; institutional management; operational & maintenance activities; statutes; human resources; water uses; infrastructure; statistics; financing; demographics
Statistics	Demographically-sorted integrated data of natural & artificial water cycles; main activities of latter
Public Information	Selected daily hydrometric data; statistics; management; legislation; main user groups; main activities

4. Data Sources

The District Water Authority oversees the daily operations of hydrological and hydrometeorological data which are reported to various users within the DWA and to VITUKI. The normal frequency of reporting is four times daily in the following volumes: domestic meteorological data are 300 KB, other meteorological are 1800 KB, and hydrological data are 700 KB. In addition to its data inputs, VITUKI sends flood forecasts and reports to the National and District Water Authorities from a set of domestic hydrological stations which monitor water more frequently — two to twelve times a day during the flood season, depending on the seriousness of the alert.

Data on the groundwater regimes and related water uses are provided much less frequently, sometimes quarterly, other times rarely. On the other hand, water quality data are transmitted from the District Environmental Inspectorates to the District Water Authorities on a monthly basis. Statistical data collected by the National Statistical Institute are made available and purchased annually.

Data concerning water infrastructure and water use have been collected only recently. They are processed by each of the water units separately. Data on domestic and industrial water supplies are included in the statistics produced by the National Statistical Institute. River and channel bed data are collected in two operations independent of one another. One operation is focused on small rivers and drainage/irrigation channels, and another collects data on rivers with flood protection levees.

5. Water Management Information System: General Principles

- Data projects and evaluation programs — e.g., databases, GIS, electronic communications, flood protection — are interdependent and therefore must be undertaken only after lines of cooperation and coordination are settled.

- Hardware and software development should be determined on the basis of the study, currently under preparation, entitled "Alternatives in the Development of an Informational Infrastructure for Hungarian Water Management."

- The nation-wide internal and external communications system must be uniform and standardized in its applications, definitions and nomenclature.

- The structure of the new data and information system must be built on the foundation of a familiar, well tested database network which makes data easily accessible; data must be open and payments that impede accessibility must not be demanded.

- A new information system should be compatible with standard relational, multi-platform software.

- The implementation of the data development plan should follow a recursive cycle of planning/programming/running; at and of the first cycle, the pilot model should be tested in one of the operational units.

- In nation-wide applications, there should be, if possible, a GIS interface based on the ISO standard (ISO/ICE 9075:1992 (E)) compatible relational database.

- Financing of the development, maintenance, and operation of the informational tools should be made within a unified and coordinated framework.

- The effective implementation of projects is dependent upon quality training of professionals; such skills training should be integral to all planning and be a permanent part of the development of the information system.

6. Water Management Information System: Principles Governing HISS

- In the initial stages of a HISS project, a clear set of objectives and detailed terms of reference should be summarized in the strategic plan.
- During the development of a HISS project, the standard recommendations of the World Meterological Organization (WMO) should, if possible, be applied.
- A major goal of such an information system should be harmonization with the water management information systems of neighboring countries, thus promoting international data exchange agreements and conventions.
- Surface water and meteorological data sets of a HISS project should encompass the entire Danube basin.
- The realization of a HISS database project should meet any safety requirements and rules for the use of the information system that are set forth in the strategic plan.
- A HISS hydrological database project should be implemented on a unified nation-wide basis; standardized user modules will be installed at water-related institutions.

7. Key Issues and Questions

7.1. COMMUNICATIONS

There is a planned bifurcated approach: urgent data will be transmitted daily by the Lotus Notes Mail System while data that are not urgent, for example, inventory data, will be sent by RDBMS replication system; international communications will be carried by telex. (A recent pilot study of the MS Access RDBMS indicated that a RDBMS with higher capability would be needed. A follow-up study revealed preference for the SQL or Oracle 7 servers. This has resulted in a decision to undertake a comparative analysis of the best RDBMS.)

In the first phase of implementation, communications will go through telephone lines. Plans call for communications between DWAs and the NWA to be transmitted by means of a special microwave channel system devoted exclusively to water management. This microwave system is already installed along the Hungarian stem of the Danube and will eventually extend to other regions of the country. Between VITUKI and the NWA there is already functioning a dedicated telephone line.

Public information will be accessed by means of the Internet through a Notes Domino server. For reasons of security, Internet access will be separated from the local LAN systems.

7.2. OPENNESS

This particular design of the communications system derives from both international agreements and the recommendations of the WMO with a view to complying with the principle of open access to data, free of charges. However, this principle cannot be fully honored until there is secure funding for the development and maintenance of the information system. Consequently, during the first phase of development, only the most general information will be free of charge.

7.3. REPLICATION OR SPATIAL SHARING?

In earlier planning, there was a general consensus that a spatially shared distributed database system should be created, wherein each local data manager would receive only part of the data relevant to his functions, and VITUKI would be given the responsibility of running the integrated information system. However, good experiences with the Notes replication changed the bias toward using replication which would provide all those who had need of data with access to the entire information system. Replication also would solve the problem of data consistency and security. The only problem remaining was when the public is given access over the Internet, that data would be vulnerable to the whims of computer hackers and those who would use the data illegally.

7.4. REAL-TIME OR DISCRETE SYSTEM?

The issue of rapid distribution of data is related to the problem of security. Various kinds of hydrological data such as those concerning protective measures or vital analyses must, of necessity, be made available very quickly. Notes mail, which is compressed, assures fast, inexpensive data transfer, while replicated data cannot move faster than the slowest part of the local network.

8. Design of a Future Data System for Hungary

At the center of the proposed system is the hydrological subsystem together with the resources inventory subsystem. These modular units provide data input, do the primary data checking and quality assessment, and store the data in an appropriated sub-database. In general, this central module will not be responsible for application-oriented data selections, reporting, etc.; rather, it assures that data is immediately accessible to authorized users. User access for most applications will be through a GIS interface.

The central hydrologigal database will consist of three main parts: the archive, the SQL database, and the operational database. The operational database will function as a bridge between Lotus Notes, Protection ISs, and the Operational Hydrological Module, and it will serve Notes oriented protection and other operational applications. Internal data storage in the HISS performs the whole process of data checking and correction, and providing tabular formats of verified data. Users will then need only to access the verified data tables.

9. GIS Interface

Most data that are in a managed system share three attributes: they have a value, a time and date, and a geocode. Data that is made more user friendly will normally be based on a GIS platform. There are two basic kinds of GIS software in general use, one is a very fast map-manager called HYDRA and the other is a dynamic, user progammable software called maGISter. Both use the same digital mapping.

HYDRA applications are Windows-compatible, pre-programmed tools, which can be used for any SQL-compatible relational descriptional database. This software does not require frequent, expensive hardware upgrades and is very good for standard, regular use. MaGISter has its own LISP program module that works for actual and dynamic selections and special topological procedures. This software is supplemented with a family of different ODBC drivers able to connect to the SQL descriptive databases too. It has a digitaliser module, which is useful for producing non-standard, high-seated maps.

The planning and development of GIS user interface for specific applications are the task of application development. The pilot HISS project will do only some standard GIS reports and use a general menu-group for data access and for the most frequently used categorics. Parallel to the HISS development project, a GIS development project which uses both the above referenced software applications was installed at the end of 1997.

SOIL VULNERABILITY MAPPING IN CENTRAL AND EASTERN EUROPE

Issues of data acquisition, quality control and sharing

N. H. BATJES
International Soil Reference and Information Centre (ISRIC)
P.O. Box 353, 6700 AJ Wageningen, NETHERLANDS

Abstract

A computerized Environmental Information System is under development for mapping the status of soil degradation and the vulnerability of soils to pollution. Geographically, the SOVEUR project is limited to 13 countries in Central and Eastern Europe. The project will: (a) develop a uniform soil and terrain database for the region; (b) provide an inventory of the current status of soil degradation, with particular attention to soil pollution; and (c) elaborate a procedure for mapping areas at risk from delayed occurrences of pollution as triggered, for instance, by acid deposition or changes in land use. Many of the required inputs, including those derived from ancillary models, will originate from disciplines other than soil science, pointing to the need for an inter-disciplinary approach to management of data on environmental resources for sustainable development. Inaccurate information may lead to conclusions and measures which endanger human health and biodiversity, which will require high costs to remedy. Thus, issues of data acquisition, quality control and sharing are critical for effective implementation of the SOVEUR project.

Key Words

data compatibility, data quality, data sharing, digital databases, environmental degradation, Europe, soil degradation, soil pollution, soil vulnerability

T. Naff (ed.),
Data Sharing for International Water Resource Management: Eastern Europe, Russia and the CIS, 187–206.

1. Introduction

As the 21st century approaches, soil degradation is no longer restricted to isolated incidents and locations; together with global climate change and ozone depletion it has become a general and contentious problem [32]. The long history of agriculture and settlement, widespread (mis)management of soils, and the intensive mining of lignite and minerals have caused a complex pattern of soil degradation and pollution all over the world [4, 29, 48]. The declining quantity and quality of water resources is also an important issue, notably for irrigated agriculture and biodiversity [1, 69]. Improved accessibility to good quality data and a sound understanding of the main processes of environmental degradation and their controlling factors, are critical for effective management of terrestrial and water resources.

Historical differences in development policies between Eastern and Western Europe have lead to differing environmental strategies [3]. The social and economic structures of many countries in Central and Eastern Europe have changed drastically during the last 50 years through rapid expansion of heavy industries, the energy and transport sector, increased urbanization, and intensification of agricultural production. Many of these changes occurred without due consideration for environmental side-effects. Thus, irrespective of the ongoing international efforts to reduce emissions at the source of pollution, there still is a vast legacy of “accumulated loads” of pollutants in soils and sediments. These chemicals may become remobilized in a biotioxic form upon climatic and land use changes [53, 54].

Estimates of areas of land affected by specific processes of soil degradation in Europe have been presented by Oldeman *et al.* [48] and revised by Van Lynden [60]. Ter Meulen-Smidt [56] reviewed regional differences in inputs and distribution of contaminants in Europe, describing broad regional trends in chemical loads. Csikos [20] reviewed important “hot-spots” of pollution in selected Central and Eastern European countries. These studies show that regional and continental estimates of the actual extent of degraded and polluted land, and of areas at risk, remain open to improvement owing in part to differences in procedures and threshold levels used [52, 68].

In addition, the establishment of monitoring networks is seen as a pre-requisite to any further coordinated approach to soil protection [60, 64, 68]. It is in this overall context that the Food and Agriculture Organization (FAO) of the United Nations and the International Soil Reference and Information Centre (ISRIC) initiated the project for the Mapping of Soil and Terrain Vulnerability in Central and Eastern Europe (SOVEUR). It is a sequel to a workshop organized by ISRIC within the framework of the Chemical Time Bombs (CTB) project [5].

2. The Objectives of SOVEUR

The purpose of the SOVEUR project is twofold. First, it is intended as a means for collating soil and terrain data for a geo-referenced assessment of the status of soil degradation, and an assessment of soil vulnerability to pollution for 13 countries in Central and Eastern Europe. These countries are: Belarus, Bulgaria, Czech Republic, Estonia, Hungary, Latvia, Lithuania, Moldova, Poland, Romania, Russia (west of the Urals), Slovak Republic and Ukraine.

Second, it is designed to produce results which are of practical utility in the geographic quantification of selected environmental issues for regional planning purposes by the countries concerned, and by international bodies such as FAO and UNEP, the United Nations Environment Programme. The data collation and analysis challenges which the SOVEUR project will have to address are comparable with the problems of planning for many regional and global systems [35, 38, 47, 70].

The focus in this study is on the soil vulnerability mapping component of the SOVEUR project. First, the broad conceptual framework is outlined. This review will serve to identify the type of data needed for the SOVEUR project, leading to a discussion of issues of data acquisition, quality control and sharing. The paper then identifies some of the anticipated problems, particularly the practical limitations of data availability and the difficulty of harmonizing data derived from disparate sources covering many inter-related thematic areas. Finally, conclusions are drawn.

3. Conceptual Framework

3.1. POLLUTION SCENARIOS

Pollution scenario identification is the first and perhaps most important step in developing guidelines for soil vulnerability mapping. Figure 1 illustrates how changes in the environmental and societal setting (politics, economics, demography) impact on human activities, and how responses to these impacts influence the environment and society. Under similar conditions of climate, soils and demographic pressures, different impacts and responses can be obtained depending on economic and political circumstances. There remains much uncertainty in the possible impacts of political, technological and socio-economic developments in Central and Eastern Europe on land use change and environmental degradation during the next decades [9, 32].

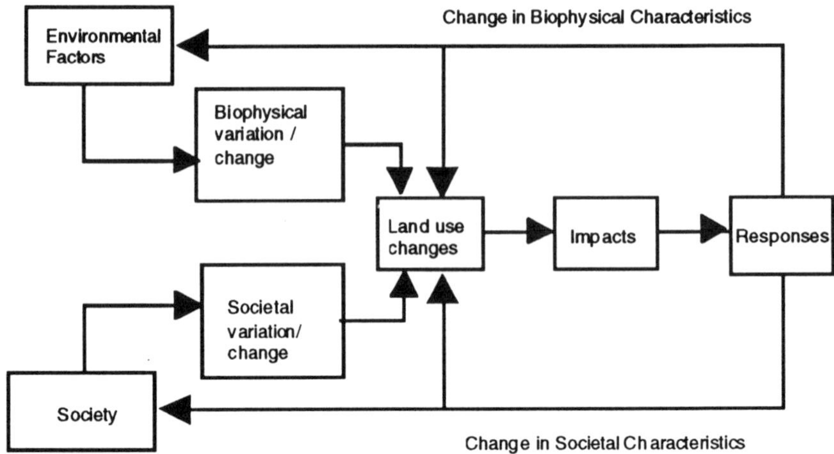

Figure 1. Schematic representation of interactive impact model, with feedback (After Kates [36]).

Each pollution scenario should consider the *source* and *type* of the contaminant, the *processes* that affect its mobility and toxicity, the *pathways* through which a contaminant and its derivatives are re-distributed in the environment, and finally the *targets* at risk from these compounds [21, 41, 51].

Alloway and Ayres [2] reviewed the major sources of pollutants in relation to the environmental media of air, soil and water with which they are transported or in which they reside. They include: agricultural activities; chemical and electronic industries; derelict gas work sites; electricity generation; metalliferous mining and smelting and metallurgical industries; general urban and industrial sources; waste disposal; transport; incidental sources, such as nuclear accidents; and long range atmospheric transport.

Six general kinds of contaminants commonly reach the soil: pesticides of varying chemical composition and reactivity; inorganic pollutants, including cationic (e.g., Pb, Zn, Cu, Cd) heavy metals and other xenobiotics such as Fluorine and Arsenicum; organic wastes, such as those arising from food processing, feedlots and municipal and industrial activities; salts; radionuclides; and acid rain [10]. Notoriously dangerous substances considered of primary concern are shown on lists, such as the EC Dangerous Substances Directive [23]. These substances may occur as solids, gases or liquids, and their source may be either point (local) or non-point (diffuse) [11].

The characteristic element of a point-source is the high local level of toxicity or corrosivity that can result in acute damage to plants, animals and inanimate structures. Generally, this type of pollution can only be mapped adequately at a local scale.

Diffuse pollution associated with atmospheric deposition and agricultural activities is ubiquitous and may spread across vast areas. An increasingly important diffuse source is runoff from paved urban areas to surface waters [50]. Generally, deposition of diffuse pollutants occurs in low concentrations, but in the longer term critical loads from these elements can be accumulated in very high levels [18, 22, 46]. A build-up to a critical mass of such pollutants may trigger the sudden release of contaminants thought to be held firmly in the soil. These pollution events have been described metaphorically as "chemical time bombs" [53, 54]. The timing and severity of the repercussions of such delayed-releases as infiltration, run-off or emissions is strongly determined by the nature and intensity of the controlling physico-chemical processes as well as the rate of decomposition and toxicity for biota of the chemicals involved [17, 62]. Being of a transboundary nature, the effects of diffuse pollutants on soil quality can best be mapped at a regional or continental level [15, 22, 27].

3.2. SOIL VULNERABILITY WITH RESPECT TO CHEMICAL TIME BOMBS

The concepts of soil sensitivity, susceptibility and vulnerability are often used interchangeably in the literature, forming the basis for confusion. The initial SOVEUR workshop defined vulnerability, with respect to "chemical time bombs" (CTB), as the "capability for the soil system to be harmed in one or more of its ecological functions" [5]. The agreed-upon ecological functions are production of humus, filtering, storage, buffering and transformations of heavy metals, persistent organic chemicals and other xenobiotics. Additionally, the protective role of soil and the genetic reserve of its flora and fauna were regarded as significant issues. Important processes (triggers) influencing the capability of a soil to hold various contaminants and pollutants include: acid precipitation; liming of agricultural land; eutrophication; salinization; water erosion; changes in climate, hydrological conditions and land use [30, 54].

Each soil is a chemically and biologically complex medium comprising weathered and newly formed mineral fragments, organic matter in various stages of decomposition, (micro)organisms, and solutes and gases in its pores. Soils in Europe are diverse, ranging from shallow and stony Lithosols to poorly drained Histosols rich in organic matter [24, 33]. Depending on their inherent properties such as content of clay, organic matter and calcium carbonate and on their cation exchange capacity, each soil will react in different ways to similar pollution and environmental changes. Thus regional differences in static and dynamic soil properties — both horizontally and vertically — will largely control a soil's capacity to control movement of pollutants. As such, each soil may be viewed as a chromatographic column, or system of "geochemical barriers" with respect to contaminant behavior [27].

Table 1 lists seven important "capacity controlling properties" affecting soil buffering and maximum retention capacity for heavy metals and persistent organic chemicals. In case of other contaminants, different sets of capacity controlling properties may have to be considered. Any environmental change that alters these capacity controlling properties will subsequently affect the maximum "sink capacity" either favorably or unfavorably [30].

TABLE 1. Important soil capacity-controlling properties for heavy metals and persistent organic chemicals

Capacity controlling property	Possible detrimental environmental effects
Cation- or anion exchange capacity (CEC resp. AEC)	Soil having a low CEC or AEC has a low capacity to retain cations, such as heavy metals, or anions, such as Arsenicum and organic anions, by sorption. The size of CEC and AEC mainly depends on clay content and type, organic matter content and type, and soil pH.
Soil reaction (pH)	A decrease in soil pH (generally) increases heavy-metal solubility, decreases CEC, and alters microbial composition and activity
Redox potential (Eh)	A decrease in redox potential (more reducing moisture conditions), dissolves iron and manganese oxides, which mobilizes oxide-sorbed toxic chemicals. Increasing redox potential mobilizes heavy metals by dissolving metal sulfides.
Organic matter content/quality (OM)	Decreasing OM content reduces CEC, soil pH buffering capacity, the sorption capacity for toxic organic compounds, soil water-holding capacity, alters physical structure (e.g., increases soil erodibility and compaction hazard), and decreases microbial activity.

TABLE 1. [Continued]

Soil structure	Altering soil structure can reduce drainage and thereby increase redox potential (more oxidizing conditions), increase soil erodibility, affect the rate of chemical release to drainage water, and alter soil pH.
Salinity	Increasing salinity solubilizes toxic chemicals by altering the ion-exchange equilibrium, increasing soluble complexation, and decreasing chemical thermodynamic activities in solution. Further, it can decrease microbial activity.
Microbial activity	Altering microbial activity and population ecology can reduce toxic degradation of organics (increase in toxic build-up), and alter redox potential and pH.

The scenario depicted in Figure 2 illustrates that with progressing acidification there may be a time when the critical load for a heavy metal is exceeded, turning the soil in to a supplier rather than a sink for this potentially harmful metal (Figure 2).

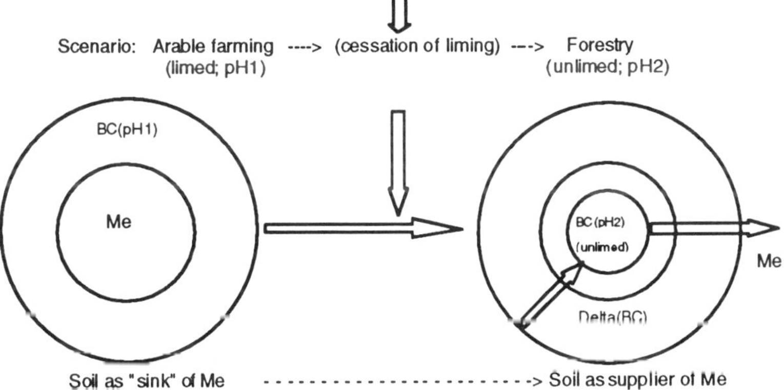

Figure 2. Schematic representation of the effect of cessation of liming, associated with a shift from arable cultivation to forestry, and acid deposition on the buffering capacity of the soil, and subsequent mobilization of heavy metals (Me).
$[pH_1 \_ 6.5; pH_2 \ll pH_1]$; BC is the buffering capacity with $BC(pH_1) > BC(pH_2)$; Modified after Stigliani [53]

A "strong delayed response" may be observed in soils which can store large amounts of potentially mobilizable chemical compounds; these sudden occurrences of pollution are the "chemical time bombs" [53, 54]. Alternatively, so-called "weak rapid responses" or "whimpers" occur where compounds are easily lost in continual small amounts. They are (shortly) delayed responses with as yet no trigger effect. These are not chemical time bombs but rather what the SOVEUR workshop participants termed "more or less" polluted areas.

In the CTB sense, the most vulnerable soils are those with high but finite capacities for storage of potentially harmful and mobilizable chemicals [5, 54]. In addition, "time-delayed" and then "sudden" non-linear releases of pollutants are important. Consequently, the chemicals of concern with respect to CTB occurrences are the long-lived species most resistant to chemical decomposition, especially heavy metals and persistent organic-chemicals.

In summary, the chemical time bomb (CTB) concept stresses: (1) the (changing) capacity of the soil reservoir to hold or release contaminants and (2) a trigger system.[5]

The severity, nature and timing of the impacts resulting from CTBs vary with:

(1) the degree of *loading* of the soil with a particular chemical
(2) the capacity or *propensity* of the soil to retain this chemical
(3) the type and intensity of the *triggers*;
(4) the *sensitivity* of individual soils to the respective triggers
(5) the *targets* affected by the released pollutants.

The combined interpretation of items (2), (3) and (4) will permit mapping of the *relative* vulnerability of soils to a pre-defined pollution scenario. In a Geographic Information System (GIS), the *relative* vulnerability map can be overlaid onto a map of current (accumulated) loadings (item 1) to show where chemical time bombs are prone to occur. At the (sub)continental level, such preliminary analyses can provide the basis for identifying areas at risk where more detailed studies of soil pollution would be useful (Figure 3). These can be used for identification of hot-spots, for case studies and remediation at the local level.

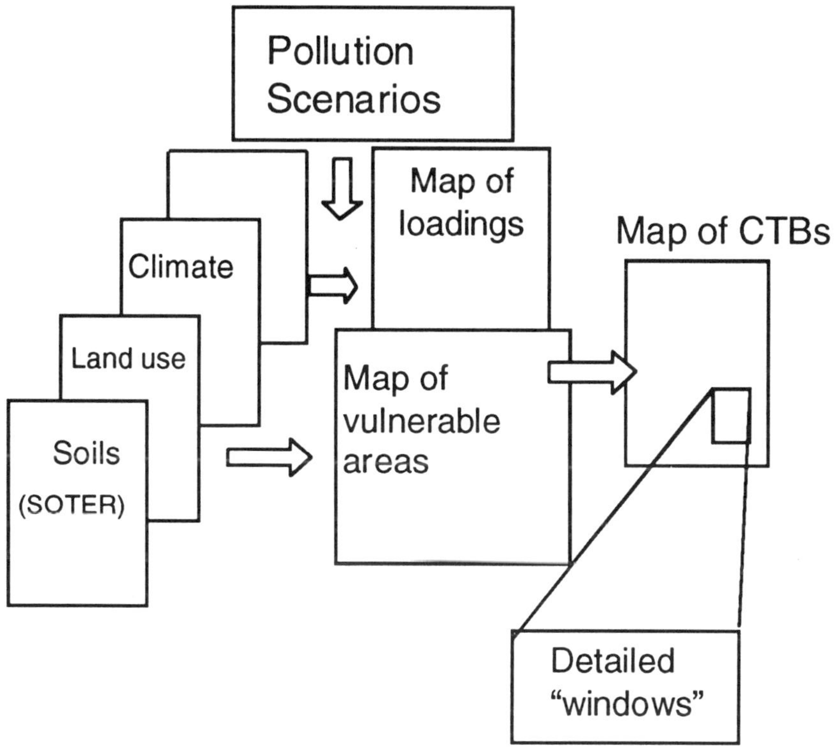

Figure 3. Schematic approach to identifying soils at risk from chemical time bombs using GIS

4. Database Issues and Constraints

4.1. USER GROUPS

As stated, the general objective of the SOVEUR project is to increase the capability to deliver accurate, timely and useful information about soils and terrain, and their inferred resilience to defined processes of land degradation. The principal users of the latter information will be staff of national environmental agencies as well as international bodies such as the Food and Agriculture Organization (FAO) and the United Nations Environment Programme (UNEP). Other users may be seen as the European Soil Bureau (ESB) and International Geosphere Biosphere Programme (IGBP), as well as individual scientists who require basic primary soil, geographic, and attributable data for a range of environmental studies. Complementary to these objectives is the need to publicize the database development activities to contributors, users and collaborating partners. This will largely be achieved through presentations at international meetings and seminars.

4.2. LOCATING THE DATA

The study and estimation of pollutant accumulation and migration under changing environmental conditions is complex. It will be necessary to bring together experts from various disciplines to solve the different pieces of the puzzle through a concerted dialogue. Relevant information will include data about climatic conditions, soil and terrain types, hydrology, and contaminant behavior in relation to changing environmental conditions. Confirming the existence and location of such data, and identifying their content, will be one of the first steps in accessing, processing and analyzing the data. Meta-databases will be key in locating the available data and their accessibility, in addition to traditional libraries[44].

4.3. DATA ACCESS AND SHARING

Data isolated from the users are of little value. Most global and many regional data sets are available on request, albeit sometimes at a nominal cost or upon prior approval from the originating body before distribution[25, 38, 58, 63]. Access to national and sub-national datasets, however, often remains subject to the wishes of the countries concerned or are subject to certain restrictions arising from their developers and contributors [28, 34].

New intellectual property laws, as proposed by the World Intellectual Property Organization (WIPO), may have potential adverse effects on the conduct of science and education. Although the WIPO treaty was withdrawn in 1996, further consideration of a database treaty is scheduled in late 1997[66]. If implemented, the International Council of Scientific Unions (ICSU) fears all sorts of compilations of data or datasets that have traditionally been in the public domain for lack of sufficient "originality" (to make them copyrightable) would now be protected against unauthorized uses. In turn, this would enable the database owners to charge whatever price they choose for authorized use [66].

Access to data may raise many issues. Some of these are technical, while others are legal, political, institutional or administrative in nature [34, 58]. A possible complication for SOVEUR are the distribution and copyright rules attached to the European Soil Database for Central Europe [33]. Due to these restrictions, the available digital soil data may not be accessible for incorporation and use in the SOVEUR project. Consequently, our national partners from Bulgaria, Hungary, Czech Republic, Slovakia and Romania may have to re-collate similar soil data for inclusion in

SOVEUR. This would entail an unnecessary duplication of effort, and inefficient use of manpower and financial resources.

4.4. DATA COLLATION AND HARMONIZATION

Effectively combining soil (degradation) information from different countries requires knowledge of sampling standards, measurement techniques, classification systems, quality assurance methods, and terminology used in obtaining and describing the information. Traditionally most soil data have been gathered to satisfy specific purposes, such as soil conservation or an assessment of land suitability in relation to crop production potentials. They may have been presented at different spatial and temporal scales. As such, the available data are not necessarily applicable in their original form. They may need to be *harmonized* for use in environmental assessments. Thus it is important that the original associated data, for example on soil analytical methods used and sources of data, be maintained and documented in the SOVEUR database for use during subsequent analyses.

4.5. SOIL AND TERRAIN DATA

Contacts, incorporating standardized guidelines, have been drawn up with institutes from the 13 participating countries for the collation of data on soil and terrain conditions as well as soil degradation status. During an initial workshop on these issues, national representatives were familiarized with the overall project goals and uniform specifications for soil and terrain (degradation) data description and handling [6]. Subsequently, they will apply the standardized methodological guidelines in their respective countries under overall coordination by ISRIC. Follow-up visits are planned to discuss possible constraints as they arise.

The procedures and data coding formats adopted for the collation of the soil and terrain data are compatible with those of SOTER, the World Soil and Terrain Database [59]. An important consideration in using the SOTER methodology has been that the resultant soil database will be suitable for multiple uses, including assessments of crop productivity and soil gaseous emission potentials [47]. In Hungary, the SOTER methodology has already been applied, at scale of 1:500,000, as a tool for evaluating the susceptibility of soils to various degradation processes [65]. The 1:2.5 million scale SOTER database for Central and Eastern Europe, which will be developed in the framework of the SOVEUR project, can later be merged into a 1:5 million scale SOTER. Upon its completion, this global database is scheduled to replace the digital Soil Map of the World [25].

Based on past experience, it is to be expected that the response rate of the collaborating partners will vary, depending largely on the amount of available information, by country, on various topics of interest collected by SOVEUR[7, 48, 61]. In some countries, there also may be policies that restrict the distribution of certain types of data on the basis of confidentiality — for instance information on radionuclides. In other cases, some of the soil attributes necessary for environmental assessment, such as soil hydraulic parameters, may not have been recorded on a routine basis. A practical solution, then, would be to infer these missing values from measured data using pedotransfer functions pending their progressive substitution by measured data[8, 12].

4.6. ACCUMULATED POLLUTANT LOADS

Accurate data on accumulated pollutant loads and the type and intensity of the environmental triggers, which may lead to toxicant mobilization, are critical for the assessment of soil vulnerability. Prieler *et al.* illustrated the importance of regional differences and changes over time with respect to total loads of accumulated heavy metals in the soil and the share of agricultural or atmospheric load in this total[49]. Ideally, data on accumulated loads (by contaminant) should be derived from country-wide monitoring programs as has been the case in Poland [55]. Alternatively, agricultural loads, by contaminant, may have to be derived from simplified mass balances, requiring access to a wide range of data [26, 49]. Data on atmospheric deposits would then have to be derived from available databases [22, 63].

4.7. DATA AND DECISION-RULE UNCERTAINTY

Once GIS data is entered into a database, GIS operations will be performed under the assumption that all data are accurate even though it is known that is not the case. This prevalence of uncertainty is a concept that many find difficult to accept [35]. The issue of uncertainty in digital databases has received considerable attention in the literature [13, 28, 37]. Lack of information on data quality, and lack of quality of the primary data itself, forms a major impediment to data sharing and use in risk assessment in soil and other environmental interpretations. Furthermore, there is uncertainty associated with the decision-rules and data formats.

Ideally, questions of necessary accuracy and precision should be determined by the uses to which the data are to be put [7, 16, 28]. However, the SOVEUR database of will, of necessity, be based on available data, each with their own characteristics of precision and accuracy. Some of the data received from member states will be incomplete or may not cover the whole territory so that important gaps may exist [52]. At times, the accuracy of the original information may be questionable. This illustrates the importance of critical checks of all data prior to their entry into the computer or GIS.

Observational data are almost always subject to error, but spatial data seem to suffer more from imperfect quality than do other kinds of data [14, 28, 45]. Common sources of error in spatial data have been reviewed by Burrough and Goodchild [13,28]. Cofino discusses the need for Good Laboratory Practice (GLP) for the production of quality information for environmental research[16]. Scale, obviously, is also important when considering auxiliary data layers, as conclusions regarding a spatial process depends to some extent on the detail and resolution of the data sets employed [31, 40, 67].

Apart from technical problems of data format conversion, it is often difficult to combine or compare data sets from different sources. Atmospheric deposition data, for example, often will be presented on a grid basis, as opposed to data on agricultural inputs/loads that will have to be generated by administrative region, and soil data that will be presented by landscape unit. Analysis and overlay of such diverse data formats in GIS will lead to "slivers." As such, uncertainties related to model and data errors are prone to be significant at the considered scale and results will mainly be applicable to large areas as a whole[42, 49] .

Decision-rule uncertainty is that which arises from the manner in which criteria are combined and evaluated to reach a decision [13, 28]. A simple form of decision-rule uncertainty is that which relates to parameters or thresholds used in the decision rule; for example, critical loads for acid deposition [43]. Different interpretations of certain qualitative concepts, e.g., slight versus strong soil degradation by water erosion, may lead to different interpretations of a given decision rule. Such differences in interpretation can be associated with different cultural, societal or political perceptions of a certain type of land degradation [41].

A more complex issue is that which relates to the structure or formulation of the decision-rule or model itself [28, 31]. A variety of theoretical approaches have been developed to accommodate for this type of uncertainty, including Bayesian Probability Theory and Fuzzy Set Theory [13, 14, 28].

5. Conclusions

The SOVEUR project will enhance scientific cooperation between European countries, notably on issues of soil database development, soil degradation and pollution [6]. An important, expected result from an operational SOVEUR database is the valuable resource information service which will be made available to a variety of agencies and organizations in national and regional resource planning. In this context, the need for good quality data and information in a uniform and user-friendly format for studies of environmental change cannot be overemphasized. This will require a holistic approach to data quality, starting with an assessment of the information needs and extending to the presentation of information to the user community.

Quality control is a major issue in spatial and point data handling, particularly when disparate sources are used; some of the "best" available data may be patchy and of uncertain quality. Uncertainties related to model and data errors are prone to be significant at the considered scale of 1:2,500,000 [42, 49]. As such, results will be mainly applicable to large areas as a whole, increasing awareness of (adverse) effects of human intervention on the quality of soil resources in Central and Eastern Europe. Generalized information on possible future regional impacts of contaminants in soils is important for policy making.

In order to model, that is, determine the outcome of future policy decisions it will be increasingly necessary to know how soil resources respond to different environmental and societal pressures, as illustrated earlier in the interactive impact-model. This will require linkage of soil databases with environmental monitoring of changes [64]. Additional research is needed on the characterization of methods, including studies of scaling phenomena and how to combine and link scales, analytical methods, methods of comparison between data sets, and linkages to models [19, 40]. As has been observed by Goodchild, "models alone cannot provide adequate solutions, and primary data do not convey absolute truth"[28].

Spatial and non-spatial databases are costly in terms of development and data collection, implying that unnecessary duplication should be avoided. Data sharing should be encouraged by good metadata and network information resource tools, which provide browsing, searching, and management capabilities for information distributed though computer networks [44]. Issues of data accessibility, copyright and legal responsibility are likely to become of increasing importance in the near future, and there is a pressing need to clarify the current situation in many of these areas [34, 66].

Finally, it can be observed that soil protection requires an integrated approach within the larger context of sustainable development, as defined by UNCED, which takes into account socio-economic variables[57]. Conservation methodologies, policy and legal instruments are needed to halt and reverse the current trend of soil degradation. A principal goal now and in the future should be to shift from reactive to preventive management of the "slowly" renewable natural resource which the soil is.

Acknowledgements

The project on Mapping of Soil and Terrain Vulnerability in Central and Eastern Europe (SOVEUR) is being coordinated by the International Soil Reference and Information Centre (ISRIC) in the framework of the Cooperative Programme of the Food and

Agriculture Organization of the United Nations (FAO) and Netherlands Government [Project: GCP/RER/007/NET].

Notes

1. Abramovitz, J.N. (1996) *Imperiled waters, impoverished future: the decline of freshwater ecosystems*, Worldwatch Paper 128, World Watch Institute, Washington.
2. Alloway, B.J. and Ayres, D.C. (1993) *Chemical Principles of Environmental Pollution*, Blackie Academic & Professional, Glasgow.
3. Appelgren, B.G. and Burchi, S. (1993) Technical, policy and legal aspects of chemical time bombs with emphasis on the institutional action required in Eastern Europe, *Land Degradation & Rehabilitation* **1**, 437-440.
4. Barth, H. and L'Hermite, P. (eds.) (1987) *Scientific Basis for Soil Protection in the European Community*, Elsevier Applied Science, London.
5. Batjes, N.H. and Bridges, E.M. (1993) Soil vulnerability to pollution in Europe, *Soil Use and Management* **9**, 25-29.
6. Batjes, N.H. and Bridges, E.M. (eds.) (1997) Implementation of a soil degradation and vulnerability database for Central and Eastern Europe, *Proceedings*, International Soil Reference and Information Centre, Wageningen.
7. Batjes, N.H., Bridges, E.M. and Nachtergaele, F.O. (1995) World Inventory of Soil Emission potentials: development of a global soil database of process-controlling factors, in S. Peng, K.T. Ingram, H.U. Neue and L.H. Ziska (eds.), *Climate Change and Rice*, Springer-Verlag, Heidelberg, 103-115.
8. Bouma, J. and van Lanen, H.A.J. (1987) Transfer functions and threshold values: from soil characteristics to land qualities, in K.J. Beek, P.A. Burrough and D.E. McCormack (eds.), *Quantified Land Evaluation Procedures*, International Institute for Aerospace Survey and Earth Sciences, Enschede, 106-110.
9. Bouma, J., Varallyay, G. and Batjes, N.H. (1997) Principal land use changes anticipated in Europe, *Agriculture, Ecosystems, and Environments* (in press).
10. Brady, N.C. (1990) *The Nature and Properties of Soils*, MacMillan, New York, 517-546.
11. Bridges, E.M. (1987) *Surveying Derelict Land*, Oxford Science Publications, Clarendon Press, Oxford.
12. Bruand, A., Duval, O., Wösten, H. and Lilly, A. (eds.) (1996) *the use of pedotransfer in soil hydrology research in europe*, Joint Research Centre/European Commission and Institut National de la Recherche Agronomique, Orléans.
13. Burrough, P.A. (1986) *Principles of Geographical Information Systems*, Oxford University Press, Oxford.

14. Burrough, P.A. (1992) Soil variability: a late 20th century view, *Soils and Fertilizers*, 56, 529-562.
15. Chadwick, M.J. and Kuylenstierna, J.C.I. (1990) *The relative sensitivity of ecosystems in Europe to acidic deposition; A preliminary assessment of the sensitivity of aquatic and terrestrial ecosystems*, Stockholm Environment Institute, University of York, York.
16. Cofino, W.P. (1995) Quality of information on contamination of the environment, in W. Salomons, U. Förstner and P. Mader (eds.), *Heavy Metals: Problems and Solution*, Springer-Verlag, Berlin, 65-78.
17. Connel, D.W. (1989) Ecotoxicology. A framework for investigations of hazardous chemicals in the environment, *Ambio* 16, 47-50.
18. Corwin, D.L. and Loague, K. (eds.) (1996) *Application of GIS to the modeling of non-point source pollutants in the vadose zone*, Special Publication No. 48, Soil Science Society of America Inc., Madison, WI
19. Cramer, W. and Fischer, A. (1997) Data requirements for global terrestrial ecosystem modeling, in B. Walker and W. Steffen (eds.), *Global Change and Terrestrial Ecosystems*, Cambridge University Press, Cambridge, 529-565.
20. Csikos, I. (1994) Langzeitumweltrisiken fur Böden, Sedimente und Grundwässer im Donau-Einzugsgebiet (Chemische Zeitbomben im Donau-Becken), in E. Heinisch, A. Kettrup and S. Wenzel-Klein (eds.), *Schadstoffatlas Osteuropa: Ökologisch-Chemische und Okotoxicologische Fallstudien über Organische Spurenstoffe und Schwermetalle in Ost-Mitteleuropa*, Ecomed Verlagsgesellschaft AG & Co., Landsberg, 317-324.
21. Desaules, A. (1991) The soil vulnerability mapping project for Europe (SOVEUR): methodological considerations with reference to conditions in Switzerland, in N.H. Batjes and E.M. Bridges (eds.), *Mapping of soil and terrain vulnerability to specified chemical compounds in Europe. Proceedings*, International Soil Reference and Information Centre, Wageningen, 23-28.
22. De Vries, W., Posch, M., Reinds, G.J. and Kamari, J. (1993) *Critical loads and their exceedance on forest soils in Europe, Report No. 58*, Winand Staring Centre (SC-DLO), Wageningen.
23. EEC (1981) *Directive on pollution caused by certain dangerous substances into the aquatic environment of the community*, Directive 76/464/EEC, European Economic Community, Brussels.
24. EEC (1985) *Soil Map of the European Communities (1:1,000,000)*, Directorate-General for Agriculture, Commission of the European Communities, Luxembourg.
25. FAO (1995) *Digital soil map of the world and derived soil properties (Version 3.5)*, Food and Agriculture Organization, Rome.

26. Fraters, D. and van Beurden, A.U.C.J. (1993) *Cadmium mobility and accumulation in soils of the European Communities, Report 481505005*, Netherlands National Institute of Public Health and Environment (RIVM), Bilthoven.
27. Glazovskaya, M.A. (1991) *Methodological guidelines for forecasting the geochemical susceptibility of soils to technogenic pollution, Technical Paper 22*, International Soil Reference and Information Centre, Wageningen.
28. Goodchild, M.F. (1994) Sharing imperfect data, in Ashbindu Singh (ed.), *Proceedings UNEP and IUFPRO international workshop in cooperation with FAO on developing large environmental databases for sustainable development*, GRID Information Series No. 22, UNEP/GRID, Sioux Falls, pp. 102—110.
29. Heinisch, E., Kettrup, A. and Wenzel-Klein, S. (eds.) (1994) *Schadstoffatlas Osteuropa: Ökologisch-Chemische und Ökotoxicologische Fallstudien über Organische Spurenstoffe und Schwermetalle in Ost-Mitteleuropa*, Ecomed Verlagsgesellschaft AG & Co., Landsberg.
30. Hesterberg, D., Stigliani, W.M. and Imeson, A.C. (1992) *Chemical time bombs: linkages to scenarios of socio-economic development, Executive Report 20*, International Institute for Applied Systems Analysis, Laxenburg.
31. Heuvelink, G.B.M. (1996) Identification of field attribute error under different models of spatial variation, *Int. J. Geographical Information Systems* 10, 921-935
32. Houghton, J.T., Jenkins, G.J. and Ephraums, J.J. (1990) *Climate Change: The IPCC Scientific Assessment*, WMO/UNEP, Cambridge University Press, Cambridge.
33. Jamagne, M., Le Bas, C., Berland, M. and Eckelmann, W. (1995) Extension of the EU database for the soils of Central and Eastern Europe, in D. King, R.J.A. Jones and A.J. Thomasson (eds.), *European Land Information Systems for Agro-Environmental Monitoring*, Joint Research Centre, European Commission, Luxembourg, pp. 85-100.
34. Jones, R.J.A. and Buckley, B. (1997) *European soil database information access and data distribution procedures*, European Soil Bureau, Space Application Institute, Directorate General Joint Research Centre, Ispra.
35. Journel, A.G. (1996). Modelling uncertainty and spatial dependence: stochastic imaging, *Int. J. Geographical Information Systems* 10, 921-935.
36. Kates, R.W. (1985) The interaction of climate and society, in R.W. Kates, J.H. Ausubel and M. Berb (eds.), *Climate Impact Assessment: Studies of the Interaction of Climate and Society*, SCOPE 27, John Wiley & Sons, Chichester.
37. Kiiveri, H.T. (1997) Assessing, representing and transmitting positional uncertainty in maps, *Int. J. Geographical Information Systems* 11, 33-52.

38. Kineman, J.J (ed.) (1993) *Global ecosystems database (ver. 1.0)*, United States Department of Commerce, National Oceanic and Atmospheric Administration, National Geophysical Data Center and World Data Center-A for Solid Earth Geophysics, Boulder, Colorado.
39. King, D., Jones, R.J.A. and Thomasson, A.J. (eds.) (1995) *European land information systems for agro-environmental monitoring*, Joint Research Centre, European Commission, Luxembourg.
40. Kirkby, M.J., Imeson, A.C., Bergkamp, G. and Cammeraat, L.H. (1996) Scaling up processes and models from the field plot to the watershed and regional areas, *J. of Soil and Water Conservation* 51, 391-396.
41. Klijn, F. (1991) Environmental susceptibility to chemicals: from processes to patterns, with special reference to mapping characteristics and spatial scales, in N.H. Batjes and E.M. Bridges (eds.), *Mapping of Soil and Terrain Vulnerability to Specified Chemical Compounds in Europe*, International Soil Reference and Information Centre, Wageningen, pp. 9-22.
42. Loague, K. and Corwin, D.L. (1996) Uncertainty in regional-scale assessments of non-point source pollutants, in D.L. Corwin and K. Loague (eds.), *Application of GIS to the modelling of non-point source pollutants in the vadose zone*, Special Publication No. 48, Soil Science Society of America Inc., Madison, pp. 131-152.
43. Løkke, H., Bak, J., Falkengren-Grerup, U., Finlay, R.D., Ilvesiemi, H. , Nygaard, P.H. and Starr, M. (1996) Critical loads of acidic deposition of forest soils: is the current approach adequate? *Ambio* 21, 510-516.
44. Medyckyj-Scott, D., Cuthbertson, M. and Newman, I. (1996) Discovering environmental data: metadatabases, network information resource tools and the GENIE system, *Int. J. Geographical Information Systems* 10, 65-84.
45. Nettleton, W.D., Hornsby, A.G., Brown, R.B. and Coleman, T.L. (eds.) (1996) *Data reliability and risk assessment in soil interpretations*, SSSA Special Publication No. 47, Soil Science Society of America, Madison.
46. Nriagu, J.O. (1990) Global metal pollution: poisoning the biosphere? *Environment* 32, 7-11 & 28-33.
47. Oldeman L.R., and van Engelen, V.W.P. (1993) A world soils and terrain digital database (SOTER) — An improved assessment of land resources, *Geoderma* 60, 309-325.
48. Oldeman, L.R., Hakkeling, R.T.A. and Sombroek, W.G. (1991) *World map of the status of human-induced land degradation: an explanatory note* (2nd rev. ed.), International Soil Reference and Information Centre, Wageningen and United Nations Environment Programme, Nairobi.
49. Prieler, S., Smal, H. , Olendrzynski, K., Anderberg, S. and Stigliani, W. (1996) *Cadmium, zinc and lead load to agricultural land in the Upper Oder and Elbe*

Basins during the period 1955-1994, Working Paper 96/30, International Institute for Applied Systems Analysis, Laxenburg.

50. Rautengarten, A. (1993) Sources of heavy metal pollution in the Rhine basin, *Land Degradation & Rehabilitation* 4, 339-349.
51. Sheppard, S.C., Gaudet, C., Sheppard, M.I., Cureton, P.M. and Wong, M.P. (1992) The development of assessment and remediation guidelines for contaminated soils, a review of the science, *Canadian Journal of Soil Science* 72, 359-394.
52. Stanners, D. and Bourdeau, P. (1995) *Europe's Environment: The Dobris Assessment*, Office for Official Publications of the European Communities, Luxembourg.
53. Stigliani, W.M., 1988. Changes in valued 'capacities' of soils and sediments as indicators of non-linear and time-delayed environmental effects, *Environmental Monitoring and Assessment* 10, 245-307.
54. Stigliani, W.M. (ed.) (1991) *Chemical Time Bombs: definition, concepts and examples*, Executive Report No. 16, International Institute for Applied Systems Analysis, Laxenburg.
55. Stuczynski, T., Terelak, H. and Dêbicki, R. (1997) The status of informational resources on soils in Poland. In: N.H. Batjes and E.M. Bridges (eds.), *Implementation of a soil degradation and vulnerability database for Central and Eastern Europe*, International Soil Reference and Information Centre, Wageningen, pp. 54-57.
56. Ter Meulen-Smidt, G.R.B. (1995) Regional differences in potentials for delayed mobilization of chemicals in Europe, in W. Salomons and W.M. Stigliani (eds.), *Biogeodynamics of Pollutants in Soils and Sediments: Risk Assessment of Delayed and Non-Linear Responses*, Springer-Verlag, Heidelberg, pp. 136-169.
57. UNCED (1992) *Agenda 21*, UNCED Secretariat, Conches.
58. UNEP (1992) *Data release policy and data archive access guidelines*, Global Resource Information Database, Information Series No. 17, United Nations Environment Programme, Nairobi.
59. Van Engelen, V.W.P. and Wen, T.T. (1995) *Global and national soils and terrain digital databases (SOTER): procedures manual*, (rev. ed.), FAO-ISSS-UNEP-ISRIC, Wageningen.
60. Van Lynden, G.W.J. (1995) European soil resources, *NATURE AND ENVIRONMENT*, No. 71, Council of Europe Press, Strasbourg.
61. Van Lynden, G.W.J. and Oldeman, L.R. (1997) *The assessment of the status of human-induced soil degradation in South and Southeast Asia*, UNEP-FAO-ISRIC, Wageningen.
62. Van Straalen, N.M. and Verkleij, J.A.C. (1991) *Handbook of ecotoxicology*, Vrije Universiteit Uitgeverij, Amsterdam (in Dutch).

63. Van Woerden, J.W., Diederiks, J. and Goldewijk, K.K. (1995) *Data management of integrated environmental assessment and modelling at RIVM*, Report 402001006, RIVM, Bilthoven.
64. Várallyay, G. (ed.) (1994) *International workshop on harmonization of soil conservation monitoring systems*, Research Institute for Soil Science and Agricultural Chemistry of the Hungarian Academy of Sciences, Budapest.
65. Várallyay, Gy., Szabó, J., Pásztor, L. and Michéli, L. (1994) SOTER (Soil and Terrain Digital Database) 1:500,000 and its application in Hungary, *Agrokémia és Talajtan* 43, 87-108 (in Hungarian)
66. Webster, F. (1997) Threat to full and open access to data, *Science International* 65, 11-12.
67. Wessman, C.A. (1992) Spatial scales and global change: bridging the gap from plots to GCM grid cells, *Annual Review of Ecological Systems* 23, 175-200.
68. Wieringa, K. (ed.) (1995) *Environment in the European Union - Report for the review of the Fifth Environmental Action Plan*, European Environment Agency, Copenhagen.
69. World Resources Institute (1994) *World Resources, 1994-1995*, Oxford University Press, Oxford.
70. Wyatt, B., Briggs, D. and Mounsey, H. (1988) CORINE: an information system of the state of the environment in the European Community, in H. Mounsey and R. Tomlinson (eds.), *Building Databases for Global Science*, Taylor & Francis, London, pp. 378-396

NATURAL RESOURCES INFORMATION

FAO's Activities in the Field of Water Resources

JEAN-MARC FAURES
Water Resources Officer
Land and Water Development Division
FAO
00100 Rome ITALY
jeanmarc.faures@fao.org

Abstract

This study is meant to inform the reader about FAO's current programs and strategies about natural resources information systems, with special reference to water resources and use.

Key Words

AEDAT, Aquastat, FAO, FAOSTAT, FRIS, GIS, GTOS, GWIS, information, systems

1. Introduction

Article 1 of FAO's constitution states that the Organization "shall collect, analyze, interpret and disseminate information related to nutrition, food and agriculture." In fact, since its inception in 1945, FAO has been constantly developing and improving its capacity to provide users with information on the state of world agriculture, fisheries, forestries, nutrition and all aspects related to natural resources development.

After the United Nations Conference on Environment and Development, which took place in Rio de Janeiro in 1992, the importance of quality information of global nature on the state of natural resources development, as well as the potential for future development and its impact on the environment, has been re-emphasized. A number of new initiatives by the United Nations and related organizations have been recently developed to address the problem better, and provide the international community with adequate information.

Of major concern is the geographical distribution of available information. In most cases, when talking of global information, it is rare to find sub-country level data.

T. Naff (ed.),
Data Sharing for International Water Resource Management: Eastern Europe, Russia and the CIS, 207–218.

Attempts to collect, process and make available data at levels below country levels have often been restricted by the difficulty and high cost of accessing the information, as well as specific methodological constraints (which are developed further in the text). Country statistics thus constitute the bulk of available information. A description of statistics available at FAO is given in the text.

The parallel development of remote sensing and geographic information systems (GIS) as tools to obtain and process distributed information at all scales have now made it possible to improve the geographical quality of information in numerous fields. Forestry, land use, topography, geology, hydrography, to mention a few, have directly benefited from this technical progress; quality coverages are now available at a global level for a series of important indicators of natural resources and agricultural development. A section of this paper presents information currently available in the FAO on GIS.

The progress made in the field of remote sensing should not, however, hide the importance of quality information at ground level. In the most fortunate cases, ground truth allows for a fair interpretation of remote sensing data which can then be used to provide distributed information at a given scale. In most cases, however, interpretation of available remote sensing data is still far from being satisfactory and a number of indicators related to natural resources development still cannot be obtained from remote sensing in a satisfactory way. These issues are discussed below with special reference to water resources.

2. Selected Natural Resources Information Systems

This section presents some of the current programs and products which are related to environmental and natural resources development information systems. Only those systems which are global by nature are described here. A list of systems developed by FAO is presented in Annex 1.

2.1. FAOSTAT — AGRICULTURAL STATISTICS DATABASE

The FAOSTAT agricultural statistics database contains time-series data, starting in 1961, for all countries and territories of the world and over 1500 variables on production and trade of primary and derived crops, livestock products, agricultural machinery, fertilizers, pesticides, land use, population, fisheries and forestry. It is FAO's reference information system for all data which is systematically collected on a yearly basis at global level.

Data are published yearly in several thematic FAO Yearbooks. The basic information is that which is included in FAO's various Yearbooks. The database is also accessible via the Internet. The software makes it possible to perform statistical operations, aggregations of countries or variables, and to download information in files which can be retrieved through spreadsheets.

The database is arranged according to 15 major domains which are described in detail in the annex. Domains which are available through the Internet for public use are: production, trade, commodity balance, population, land use, food aid, fisheries and forestries.

The *production* domain is a compilation of statistical data on basic agricultural products and related information for all countries and territories of the world. Statistics include data series on area, yield and production of numerous crops and on livestock numbers and products. The *population* domain presents an annual time series, for all countries and territories, of the following items: total population, males, females, rural, urban, agricultural and non-agricultural, and active population. The *land use* domain presents data on total country area, land area, arable land, land under permanent crops, permanent meadows and pasture, forests and woodland and other land, plus the area equipped for irrigation.

2.2. THE FAO GIS DATASET

A number of FAO activities call for the support of a geographic information system capability. FAO's GIS Unit has been collecting numerous GIS coverages related to natural resources and agriculture through various projects. The GIS Unit works on ArcInfo in the Unix environment.

The main dataset developed by FAO and used as basemap for most of FAO's global land use studies is the digital version of the FAOIUNESCO soil map of the world (SMW) at the 1:5,000,000 scale. It is the only GIS dataset which is readily available for public use. It contains two types of files: SMW map sheets and derived soil property files with images derived from the soil map of the world. Derived soil property files consist of interpretation programs and related data files. They include texture, slope, pH, organic carbon content, C/N ratio, clay mineralogy, soil depth, soil moisture storage capacity, and soil drainage classes.

A 1:10,000,000 coverage of agro-ecological zones of the developing world is also available. The agro-ecological zone (AEZ) database was originally developed as a statistical/tabular database as part of a global study of population supporting capacity in the developing world. It was subsequently developed into a spatial database on GIS and is currently being expanded to include developed countries. It contains information on soils and land form, temperature regime and length of growing period, agro-ecological zones, forests and protected areas, and land suitability for about 30 main crops. The database has been used in various applications such as the determination of the extent of potential arable land for FAO prospective studies on agriculture.

Several regional or continental data sets at scales ranging from 1:1,000, 000 to 1:33,000,000 are also available for parts of the world, with more information on Africa. This includes coverages on political and administrative boundaries, population, hydrography, physiography, geomorphology, rainfall, climate, problem soils and soil development constraints, and river basins.

A project aimed at preparing an electronic atlas which would present all the information currently available on the state of nutrition and agriculture is in preparation.

3. GTOS — Global Terrestrial Observing System

The Global Terrestrial Observing System GTOS) is a joint initiative of UNEP, UNESCO, FAO, WMO and the International Council of Scientific Unions (ICSU).

The Secretariat is provided by FAO. GTOS was established in January 1996 in response to a call for a better understanding of global changes in the Earth system. It aims at improving the quality and coverage of terrestrial ecosystem data and integrating them into a world-wide knowledge base.

GTOS is not an information system as such. Rather, its objective is to foster an integrated, equitable partnership of data providers and users at national and international levels. Eventually, the heart of GTOS would be a permanent observing system for the world's key managed and natural terrestrial ecosystems.

It is expected that when GTOS is fully operational, global geo-referenced data will be distributed electronically through a network of national and regional centers, stored in relational databases at regional and global levels, and made freely available to users through CD-ROMs or via the Internet. An important endeavor is being made to harmonize methods of taking measurement and the terminologies used should eventually increase the use and value of terrestrial ecosystem data.

GTOS forms a unified information system together with the global observing systems for climate (GCOS) and the oceans (GOOS). It is expected that GTOS would eventually help find answers to five key questions:

- What are the impacts of land use change and degradation on sustainable development, and what are the consequences in terms of global food production?
- Where, when, and by how much will demand for fresh water exceed supply?
- Where and when will toxic pollutants cause major threats to human and environmental health?
- Where and what type of biological resources are being lost?
- What are the impacts of climate change on terrestrial ecosystems?

4. Other Relevant Information Systems

4.1. AFRICOVER

For the last two years, FAO has also been involved in the organization of the Africover project. The objective of this project is to establish, for the whole of Africa, a digital geo-referenced database on land cover and a geographic reference (geodesy, toponomy, roads, hydrography) at a 1:200,000 to 1:250,000 scale. This base will further be generalized at a 1:1,000,000 scale, updated, made homogeneous, comparable and compatible with thematic and geographic points of view, for the whole continent.

The project relies mostly on a combination of remote sensing and GIS techniques. One of the main benefits expected from this project will be the harmonization of legends for land cover mapping. An international working group on classification and legend was set up with the task to define a standard classification for land cover in which special attention is given to cultivated land classification.

A further objective of the Africover project is to strengthen national and sub-regional capacities for the establishment, update and operational use of geographic reference maps, land cover maps and geodatabases.

4.2. AGROCLIMATOLOGY AND EARLY WARNING SYSTEMS

A series of databases have also been developed within the framework of FAO's early warning system. They are:

- The ARTEMIS CD-ROM is an historical database of approximately 500 megabytes of 10 day vegetation index images covering Africa over the period from 1982 to 1991

- FAOCLIM 1.2 is a database on a CD-ROM providing monthly averages and time series of agroclimatic data for about 20,000 stations worldwide. A user-friendly interface under Windows allows the data to be selected according to their location, country, type, period covered, output format, etc.

- The IGADD CPSZ (crop production systems zones) is a computerized database which provides detailed information on actual farming in the IGADD region. For each of about 1000 homogeneous areas (mostly administrative units or sub-divisions thereof), more than 500 variables were collected or calculated, covering the main crops grown, agronomic data such as crop yields, pests and diseases, phonology, livestock data, plus a number of parameters describing the physical environment (rainfall, PET, elevation, slope, typical dates of the beginning and end of the rainy season).

Beginning in 1991, the Agrometeorology Group of FAO began systematically to assemble sub-national crop statistics from African countries (AGDAT database), still as background information for crop monitoring and forecasting. The exercise is now being completed. The information will be distributed using another FAO software, the Database Map Viewer (DMV) which was specifically designed to distribute and visualize flat-file databases which contain attributes attached to geographic units, as is typically the case with agricultural statistics which cannot be handled under a classical GIS. DMV offers different projections and export facilities to the most popular GIS systems.

5. GWIS — The Global Water Information System

Agriculture takes about 70% of the world's water withdrawal, making that sector the most important user of water resources. As pressure on water resources increases and more and more regions of the world are facing water scarcity, the role of water resources in agricultural development, its potential for extension, and its competition with other sectors, are of critical importance in assessing future irrigation extension and agricultural production. It is estimated that about 17% of the world's agricultural land is under irrigation, and that irrigation represents 40% of the world's agricultural production.

Although future prospects indicate that the bulk of agricultural production increase in the future will come from irrigated land, no one is in a position to predict how increasing water scarcity will affect irrigated production. The rate of irrigation expansion has been decreasing from 2.5% per year in the 70s to 1.2% per year in the 90s and it is not expected to rise again. The exponential increase in overall water withdrawal over the century is also an indication that it will be increasingly difficult to collect water for irrigation in the future.

As the UN Agency in charge of all aspects related to agricultural production, including long term forecasting, and as a member of the UN-ACC Sub Committee on water resources, FAO is committed to provide global information on the state of water resources and use, in relation to irrigation. The Global Water Information System (GWIS) is intended to provide users from the international community with the most exact and relevant information on the state of water resources and use, with a special emphasis on rural water use, including irrigation.

GWIS was initiated in 1993. It is based on a number of principles, its main objective being to provide cost-effective, relevant, and quality information on a global basis. The underlying principle of the system is that information on water resources and use for agriculture is better known at country levels than at regional or global levels. Country statistics and specific country studies are being collected and processed to extract the required information and to develop regional and continental databases.

However, this choice has two major implications. First, there is a need to develop very detailed standards to compute the different indicators which are utilized to best represent the state of water resources and use in agriculture. Definitions of water resources, irrigation, water withdrawal etc. are all but trivial, and can be interpreted in very different ways in different countries. Thus, a set of well-described indicators has been developed from the outset of the program and is being adapted from the experience gained in progressively collecting country information.

The second issue of major importance when collecting data at country level is the role of water as a transboundary resource. In fact, not only can water resources be computed in several different ways by different countries, but also the computations of transboundary flows are often performed in different ways on both sides of the border. A physical approach, based on a hydrological division of the land, needs to be superimposed on the in-country division and there must be administrative units to ensure the integrity of water resources assessment at global level.

In order to provide the most relevant information for planners and researchers of the international community, GWIS has thus been developed in two complementary programs:

- A program, known as Aquastat, is used to collect statistics on the main indicators related to water resources and use in agriculture at country and sub-country levels.

- The development of a GIS-based hydrological and environmental capability to merge information collected from countries with currently available coverage to provide a global picture of water resources and withdrawals, based on a hydrological division of the land, i.e., river basins. This program is known as the Fresh Water Resources Information System (FRIS).

6. Aquastat — Country Statistics on Rural Water Use

The main purpose of the Aquastat program is to select systematically the most reliable information on water resources and water uses in countries and make it available, in a standard format, to users interested in global or regional perspectives.

The survey was carried out for the African continent in 1995, for the Middle East in 1996, and is on-going for the countries of former USSR and for Asia. Latin America was targeted for survey in 1998.

The surveys follow a standard methodology, which relies heavily on national capacities and expertise:

- country-based reviews of literature and existing information, from national statistics and yearbooks, water resources and irrigation master plans, FAO and other international agencies' reports, regional surveys, etc.
- data gathering through a detailed questionnaire
- standardization of available data
- data processing and critical analysis of the information, validation through data-processing software developed specifically for the survey
- preparation of country tables and of a country profile which is then submitted to national authorities for comments and possible corrections

On the basis of the country surveys, regional and continental summary tables are prepared and cross-checking is carried out on transboundary flow.

The data collected through the Aquastat survey are classified according to the following categories:

- renewable water resources
- water withdrawal by sector
- non-conventional sources of water
- origin of irrigation water
- irrigation potential
- land salinized by irrigation
- health issues related to irrigation
- wastewater production and treatment
- irrigated and rainfed crops and yields
- irrigated areas and irrigation techniques
- costs of irrigation and drainage development
- drained areas and drainage technology
- types of irrigation management methods

Additional information on irrigation development, the institutional environment, and trends in water resources management is also presented for each country.

One of the major advantages of the program is that a bibliographical reference is provided with each data set entered into the system. This allows not only for keeping track of the source of information, but also for storing data coming from different sources and to select the most reliable information on the basis of its source.

The first two regional surveys have shown that it is particularly difficult to collect systematically quantified information on a series of indicators for all countries. In general, countries facing water scarcity have produced a large number of reports on water resources and several publications are available, but they all rely on a very small number of basic data and computing methods vary from one country to another. In countries for which water resources are not an issue, information on irrigation and water resources is usually almost nonexistent.

The survey also shows that there is no agreed-upon method to compute key elements of the water balance such as gross water withdrawal for irrigation. This figure is never obtained from direct measurements. Generally, it is based on a rough assessment of crop water requirement multiplied by the extent of irrigated land, or on designed water withdrawal for large schemes. The Aquastat survey has so far used these figures, but there is still room for a standardized approach in that field.

The importance of standards in defining the different indicators used in this survey was also revealed during the first surveys. Primary indicators like arable land, cultivated land, irrigated areas, or water resources may bear very different meanings from one country to another with significant variations in global results. Although the indicators used in the survey have been tested in numerous countries, modifications are still necessary to accommodate specific regional characteristics.

7. FRIS — The Fresh Water Resources Information System

Two years ago, FRIS began to take advantage of the recent technical advances in the application of Geographic Information Systems (GIS) to hydrological modeling. The FRIS-GIS combination is designed, as a complement to the Aquastat program, primarily to address FAO's internal requirements for regional water resources assessment. GIS is combined with several hydrologic models to provide a geo-referenced evaluation of water resources for large river basins. The advantage of this approach is that it allows the user to visualize the distribution of water resources inside the basin boundaries, rather than providing only estimates at point locations.

To develop and test the model, the Niger river basin, covering 2 million km^2 and a range of climatic conditions, was studied in a pilot phase. This pilot program tested the possibility of combining GIS and hydrologic modeling for regional scale river basin planning. The model uses available coverage of topography, climate, soil characteristics, vegetation, and geology. Topography is used to recreate artificially, in GIS, the river network and to provide the necessary linkages among all the elements of the basin area. A soil water balance model, combined with flood routing and groundwater models, produces a computerized map from which a variety of studies and analyses can be performed:

- assessment of water resources distribution over a region
- assessment of exchange of water among countries and administrative regions
- impact of water development scenarios (dam construction, river diversion, pumping, or land use change on the hydrological regime of the basin.

The model was primarily developed to produce regional estimates of water resources, but it soon became apparent that it could be adapted to provide a tool for the planning of water management in large river basins or at country levels. Future activities will concentrate on the preparation of guidelines and a computer package to apply the FRIS tool for river basin planning. At present, pilot applications are being developed in member countries.

7.1. EXAMPLES OF RESULTS

The Aquastat survey has now been carried out for Africa, the countries of the Near East and the countries of the former USSR. The survey has made it possible to collect and present systematic information about the state of irrigation, water resources, and water use in the countries of these regions, and to present the results at a regional level.

The Aquastat survey for Africa, using some of the techniques developed in the FRIS program, has thus been used in a study aimed at assessing irrigation potential for the continent. A publication, *Irrigation potential for Africa: A basin Approach*, (available through FA) has been prepared as a result of this study.

The combination of both the Aquastat and FRIS programs has made it possible to make the best use of the information available (through Aquastat) from the countries, and to integrate the results at the level of major river basins through GIS. The results show the degree of stress on water resources for each of the 25 river basins and illustrates possible conflicts among different regions laying claims on the same body of water.

In the central Asian countries of the former USSR, the Aquastat survey has shown that it is very difficult to obtain reliable and systematic information about the estimated natural flow of the main rivers of the region, including the Amu Darya and the Syr Darya which feed the Aral Sea. Runoff data of the region are rare and do not represent a natural situation. Most of the rivers are affected by significant withdrawals since the early 1950s. Furthermore, the agreement that exists among these countries to share the water of these two rivers are based on complex computations of river runoff and return flow from irrigation; this makes it practically impossible, except for some global figures, to know the original statistics used by the planners who prepared the initial agreement.

Finally, the region is symptomatic of the typical situation of rivers that flow from a humid to an arid area and undergo important losses in natural channels on their way to the sea. In such a situation, the resources of individual sub-basins are not additive and their losses have to be taken into account in the computation of water resources.

In this case, GIS was also used to complete the information obtained through the Aquastat survey to propose a first estimate of the distribution of water resources over the area of the different sub-basins. Runoff data for several stations of the area were collected by the Global Runoff Data Center (Koblenz, Germany) and were analyzed with a view to retaining the most reliable data sets. A map dividing the region into river basins covering an average area of 5 000 km^2 each, was prepared on the basis of a 30 inch digital elevation model and corrected with the help of the 1:5,000,000 map of the Times Atlas of the world.

A simple annual water balance model was then developed, partly on the basis of works developed by the USSR Committee of the International Hydrological Decade, and partly on the basis of the information collected through the Aquastat survey for the countries of the former USSR (area under irrigation, transfers, and water withdrawal). Reasonable results now make it possible to obtain a first estimate of the distribution and order of magnitude of river flows in the area, and to complete the information obtained from the Aquastat survey.

These two examples show the importance of an approach combining the collection of information from different sources, including countries, and the modeling of parts of the water balance, thanks to recent advances in GIS technology.

7.2. INFORMATION EXCHANGE

By mandate, FAO is committed to collect, process, and disseminate information on all aspects related to natural resources development for agriculture. An important part of its regular budget is used in maintaining networks of data exchanges at all levels and to promote the exchange of information. FAOSTAT is the main expression of this work for all country statistics and provides all users with timely and quality information for about 1500 indicators related to food and agriculture. The recent development of the FAOSTAT Internet database now allows access to all the information it contains to all users connected to the Internet. Printed copies of the FAO Yearbooks are still issued annually for those not having access to the Internet. Other thematic databases have been developed which are made available through Internet.

Simultaneously, as indicated above, FAO is developing several thematic initiatives to advance, in collaboration with partners, information systems in several fields related to national resources development. Most of these systems or databases are in an intermediate stage of development, and the information they contain is either incomplete or insufficiently validated. In such cases, FAO does not make the information available for public use until it is considered final. This rule also applies to GIS coverage; only when a GIS coverage becomes final is it made available for public use.

Appendix 1: Selected information systems of FAO

Title	Description
Aezcss	Agro-ecological land resources assessment and land use analysis
Agrimarket	Market data processing software
Agris	International cooperative bibliographic system
Apis	Agricultural policy information system
Aquastat	Database on water resources and water use in the world
Cobs	Country cereal balance system
Climwat	Climatic databasc for Cropwat
Cropwat	Irrigation planning and management tool
Ecocrop	Crop environmental requirement database
Enreq2	Software to calculate human energy requirements at country level
Faoclim	Agroclimatic database
Faosoil	Derived soil properties and soil and terrain suitability for agriculture
Faostat	FAO's main statistics database
Farmap	Farm survey data collection and analysis software
Fisat	Fish stock assessment tool
Giews	Global early warning information system
Globefish	Electronic databank on fish marketing and trade
Luse	Land use and land cover database
Pqcard	Information on pest and plant quarantine
Price index	Monthly prices for several agricultural items
Rice export	Rice export volumes by destination and origin
Sdbm	Multilingual soil database
Simis	Scheme irrigation management information system

SOFA	Database on the state of food and agriculture
Swarms	GIS for management of locust environmental data
(no name)	Global plant quarantine information system
(no name)	Microbanking system
(no name)	International directory of agricultural engineering
(no name)	Annual trade by country in cereals
(no name)	Rice export price index
(no name)	Database on maximum residue limits for pesticides in food
(no name)	Nutrition country profiles database

INTERNATIONAL ENVIRONMENTAL DATABASES: OPTIMAL UTILITY / LOW COST

IRENE LYONS MURPHY, PH. D.
Colorado State University
2005 37th Street, N.W.
Washington, D.C. 20007 USA
imurph@aol.com

Abstract

Data and information must be shared among countries and regions if increasing risks from environmental disasters are to be avoided and adequate protection provided for dwindling natural resources. The structured sharing of accurate, timely, and compatible data about the status of such resources nationally, regionally, and globally is intrinsic to their equitable use world-wide. The development and possible utility of two environmental database programs — an information system developed for the Danube River Basin in 1990-1991 and an EU Phare -European Bank for Reconstruction and Development (EBRD) sponsored Environmental Standards Database (ESD) — are traced and their potential for use in improved environmental management is evaluated. The potential value of each data system to national and international decision-makers, scientists, non-governmental organizations and the general public, as well as the cost of keeping them current for their respective uses are described.

Key Words

Danube Basin Information Network, DBIN, Danube Environment Program, DEP, Danube Information System, DIS, Danube River, environmental information systems, environmental standards database, ESD, VITUKI

1. Introduction: The Need for Optimal Sharing of Information and Data

Countries negotiating the improved environmental management of a shared natural resource need data and information at key decision points during the progression from initial discussion to final agreement. When a convention provides for the establishment of an international agency to carry out its mandate, its parties continue to need updated or additional data and information to evaluate past decisions and to help guide future choices.

T. Naff (ed.),
Data Sharing for International Water Resource Management: Eastern Europe, Russia and the CIS, 219–228.

Currently there are considerable environmental data available on the Internet. Interested parties to an agreement to improve the management of a shared natural resource have access to a wide variety of reports, including many on air, water, and soil pollutants, their impact on the quality of the natural resource under negotiation, as well as estimates of the effect of water quality on economies and social systems of each country and the region. Decision-makers can learn about methods of treatment and their costs, environmental policies advocated and practiced by national governments and international agencies, and other relevant matters.

Widely available data and information will, for the most part, be general in nature. The more specific questions needed for operational programs to resolve transboundary environmental issues are far more difficult to answer. Governments must be ready to negotiate about the specific problems which arise from such pollution, for example: What is the nature of the serious pollutants which have been identified? What does the monitoring data of each country show about levels and effects? What corrective measures are under way and do they meet the norms advocated by international authorities?

Ideally, the parties to an agreement must be ready to develop and share timely, accurate, and compatible data and information about the source and the optimal resolution of the issues under review. Additional information is needed about government contacts and how ministries share responsibilities for the issues under review; summaries of legislation affecting environmental management, including the natural resource quality standards in effect, and reports on the status of the environment.

2. Responding to Information Problems

General data about environmental phenomena and their management provide only limited help to serious, fruitful negotiations for multi-party agreements on troublesome transboundary natural resource issues. Web sites do not provide uniform, comparative data and information about national environmental policies in countries in the region. The development of relational databases of trend data about the status of potentially toxic chemical and other substances on a national and international basis has been proposed but is not yet widely established. Most countries, in fact, limit the release of data about air, water, and other media levels of a general nature to periodic summaries, if they report them at all.

Clearly, there is a need to encourage the use of new communications technology beyond limited or perfunctory releases of information. There is a particular need for developing nations that are in transition to harmonize legislative policies with practices of more developed countries with more experience and technology, including those policies concerned with improved environmental quality; the latter has significantly increased the need for a variety of legal, scientific, and technical data and information. Two databases developed during the early 1990s reflect the special interest of international funding sources as they tried to estimate environmental damage in eastern Europe and the cost of treating it.

3. DBIN and ESD

One system, the Danube Basin Information Network (DBIN), evolved from an early initiative by Danube riparians to establish a comprehensive decision support system, InfoDanube, which was developed in 1990-1992. DBIN has received support from the NATO Scientific Affairs Division Environmental Program, among other sources, including a number of specialists in the Danube countries. The second system, the Environmental Standards Database (ESD) was commissioned by the European Union Phare-EBRD (European Bank of Reconstruction and Development) in 1993 and completed in August 1994. ESD provides access to the complete environmental quality standards of 22 countries including those issued in EC Council Directives, and to the recommended criteria of the World Health Organization. Several goals were set for ESD, including aid to the global harmonization of standards, guidance for investors in the newly democratized countries, and for public and private organizations concerned with the development of waste treatment facilities and industrial plants.

The two systems, DBIN and ESD, will be reviewed from the standpoint of their respective goals, their achievements, as well as their future use and development and distribution costs.

4. International Basin Environmental Needs: The Importance of Shared Data

International agreements among river basin countries have been in effect for centuries. They have been limited mainly to the resolution of boundary or navigational issues. They have always required the sharing of some data to facilitate problem solving. Bi-lateral and multi-lateral water resource projects have been supported by transboundary pacts which provided for an on-going sharing of data, primarily about hydrologic conditions on the shared section of the river.

Comprehensive agreements in international river basins remain rare. An estimated 14 of the 240 international basins have such agreements in force. [1] Interest in promoting transboundary conventions and treaties has been positive in many parts of the world. There has been an increasing participation by all basin countries in resource development decisions, a growing interest in and affirmation of the application of international law principles, and the enactment of stronger environmental policies in many countries.

A 1986 meeting of members of international river basin commissions and other specialists, funded by the Ford Foundation and held at IIASA headquarters in Austria, cited basic requirements for improved basin management. [4] These included the adoption of a convention by all basin countries that would stipulate sharing data, and the establishment of a commission with authority and funding to carry out a number of programs. Necessary programs included adherence to a basin-wide economic development plan and the sharing of accurate environmental and other essential data and information. Existing international river basin management conventions have specified agreements to share data and information but none has yet installed a comprehensive information system.

5. A Comprehensive Information System For The Danube: An Initial Attempt

In 1990 I received a Fulbright grant to support the development of a prototype comprehensive information system for the Danube River Basin. A basic goal was to assist international and national decision-makers as they grappled with the need to improve the environmental conditions of the Basin. The design and implementation of the system received help not only from scientists and technical experts in the host organization, the Bulgarian Academy of Sciences, but from many of the other Danube countries. The system's utility was under continual evaluation during its development and many changes were made in response to recommendations.

The basic system was designed for use on PC DOS computers and did not need software. It included the following databases:

- references to bibliographic systems to assist research,
- a directory of government, non-government, and research and educational organization officials and other contacts concerned with basin natural resources,
- a legislative database which identifies major laws and summarizes them,
- water quality standards for the then eight (1990) Danube countries, and
- the yearly results of water quality monitoring of 35 parameters, including potentially toxic elements, at 11 transboundary Danube sites, authorized by the Bucharest Declaration adopted by the Danube countries in 1986.

These databases were relational, that is they shared common fields and, with the help of a compiling software such as Clipper, could develop reports on various topics. The database management system software used was Alpha 4. The grant helped to distribute it to government and research specialists in each country. Relational databases made it possible to develop reports which would, for example, compare water quality standards among countries to test their compatibility, to relate the standards to the Bucharest Declaration data, or react to increases in pollution which suggested emergency situations.

InfoDanube was used to formulate the technical section of the meeting of government and research specialists held in Sofia in the fall of 1991 with the aim of providing coordination in the improved management of the Danube Basin. Agreement at the conference led to the creation of the Environmental Program for the Danube River Basin (Danube Environment Program or DEP). [2] The Fulbright grant provided for travel to the Danube countries and consultation with water and environmental ministries, research centers, and non-governmental organizations regarding the development and use of InfoDanube. After a year at the Bulgarian Academy of Sciences, additional funding was made available by the Regional Environmental Center in Budapest to continue work, including a workshop series in Bulgaria, Romania,

Czechoslovakia, and Hungary. The system received a considerable amount of help from experts in each country.

The DEP sponsored a number of programs designed to improve management of the basin, including more accurate water quality monitoring and laboratory techniques, identification of priority sites needing clean-up in each of the former Eastern bloc countries, as well as a web site providing information about the program. Funding was not provided for a comprehensive system until near the end of the DEP in 1997. The DEP was due to be superceded by programs under the guidance of a new Commission for the Protection of the Danube River in the fall of 1998. The Convention provides for the sharing of data and information by each country.

6. An Interim Step: The Danube Basin Information Network

In May 1996, a NATO funded Advanced Research Workshop brought together government ministry representatives and other specialists in the Danube Program to investigate current data-sharing in the basin and to suggest how it might be extended to the whole basin. [3] In view of the fact that the DEP had not yet developed a comprehensive information system, the group suggested the creation of such a system. A NATO Linkage Project facilitated the development of the Danube Basin Information Network (DBIN) which has been available on a protected web site on the Internet since 1997. The Central European University Budapest (CEU) has provided a web master and a server for this purpose. In addition to CEU, a consortium of other institutions — the University of Ljubljana, Colorado State University, and the Hungarian Water Research Institute VITUKI continues to work on the DBIN.

The system is designed to serve a wide spectrum of scientists, government staff, and individuals in non-governmental groups in the public and private, profit and non-profit sectors. The relational databases, which are now using Microsoft Access, can accommodate data without the need for additional software. The design also makes it possible to answer specific questions about the status of water resources (hydrological and related environmental phenomena) in the basin and national and international management policies which affect the status of these resources.

The DBIN design as it was developed by the consortium ultimately combined two sets of databases: one to display the environmental status of the basin and the other to identify government officials and others responsible for making decisions or providing information about the management of the basin. DBIN also includes relational databases which summarize legislation and identify selected water quality standards for each country. Links with relevant international and national sources are also provided.

The DBIN design makes it possible to track the decision processes of national governments, regional and local entities, and international agencies which have water management responsibilities. These decision processes are inevitably complex for a number of reasons. Tracking them through the multiple agencies which have responsibilities for water management presents difficulties. A complete analysis would require a significant amount of research in both domestic and foreign policy spheres. DBIN an introductory guide to this process.

While considerable data and information are now included in DBIN there are, nevertheless, notable gaps. The evaluation of the availability and quality of water and other natural resources depends on the use of standardized databases within each country and the continued development of an international network of monitoring sites. The extent to which the Danube countries are willing to participate in the development of such data and to make it public has not yet been determined. The DEP sub-groups, unfortunately, have not coordinated their use of databases. Thus data collected for use in the Accident and Emergency Warning System, about hot spot priority sites, and for the Danube Trans-National Monitoring Network are not contained in a compatible database format. Making them compatible, however, is a necessity and would be a relatively simple matter.

7. A Danube Information System for the New Commission

Many features of DBIN can be helpful to the Danube Information System (DIS) which has been recommended by the DEP and is being contemplated by the new International Commission for the Protection of the Danube River. The International Commission was expected to be formally installed in the fall of 1998. In view of the general reluctance of the countries to provide data about water quality a revised design of the system has been recommended. The web site of the Rhine Commission (http://www.iksr.org) has established an interesting precedent. [See Appendix 1, Figure 1, for the recommended DIS home page.]

Note that the recommendations include a map of the basin which would provide connections to the web sites of the Danube countries. Information on the basin would include a description of its hydro-geology, eco-systems, a contact directory, the national environmental action plan, other agreements on the river, and reports on hydrologic, water quality, and other trends.

Each country would be requested to maintain a country home page with complementary databases (i.e. contact directory, water quality standards, summary of legislation) and to provide other data and information at its own discretion.

Confidential, detailed water quality and other hydrological data would be maintained separately at Commission headquarters and in each country ministry, accessible only by a password and not available to the public. (See Appendix 2, Figure 2. Among other purposes, the confidential set of databases could be used to generate reports for public consumption and to facilitate decision making by the Commission directorate regarding the need for emergency preparations and the funding of clean-up spots. The relational databases can accommodate anticipated developments in electronic communications. Interactive applications now make it possible for users to obtain answers to detailed, complex questions in a "user-friendly" mode. (See the web site for the Consortium for International Earth Sciences Information Network [http://www.ciesin.org].)

8. An International ESD: The Harmonization of Environmental Standards

The Environmental Standards Database (ESD) was prepared as part of the European Bank for Reconstruction and Development (EBRD)/European Union PHARE project, "Environmental Legislation and Standards in Western and Eastern Europe: Towards Harmonization." The project's sponsors asked that the system be user friendly, following a pattern developed by the InfoDanube program. Entering and updating data are not difficult. ESD was intended to be a flexible tool for research and for decision making, to be used in conjunction with other sources of information about environmental standards. It can respond to many needs, from identifying a single standard for a single use in a single country to reporting on all the drinking water standards for any or all 22 countries involved.

Its potential for a very large number of uses is very high. It can, for example, advance the harmonization of standards between east and west, as well as the methods of measurement now in use. It was interesting to note that only one of the 22 countries, The Netherlands, keep all of their air, water, soils and sediment standards together, publishing them in one volume biennially. It was frequently difficult to locate those in charge of various media standards in the remaining countries. Thus another use for the system was to offer a single point of access for all of the environmental standards of each country.

9. Content, Design, Distribution

The ESD project was under the guidance of EBRD and executed by a team assembled by the international consulting firm of Booz, Allen Hamilton. The standards were assembled over a nine-month period under the supervision of a project director stationed in Budapest. The Central and Eastern European and Baltic countries contributing standards included: Bulgaria, Croatia, Czech Republic, Hungary, Romania, Slovakia, Slovenia, Albania, Poland, Estonia, Latvia, and Lithuania. These Eastern European countries received special assistance for the preparation of their parts of the ESD, which included funding for training in the use of the system. The remaining countries were: Austria, Finland, France, Germany, Japan The Netherlands, Sweden, Turkey, the United Kingdom (England and Wales), and the United States.

ESD was designed for use on a personal, IBM-compatible computer and contains environmental standards for air and water quality, noise, soil, sediment, occupational health, biota, and radiation. The database identifies numerical values, and provides additional information about them in an expanded text field. It refers to document sources and the names and addresses of contact points. It can be sorted in hundreds of ways, by parameter, by media, by country. In addition, its country report section contains information on each country. ESD can also be used to develop and print reports, accompanied by graphics.

Compatibility of terminology was sometimes a problem. The sponsors of ESD had requested that all data be in English; translations of some sets of databases were required and were verified by experts. The use of chemical abstract numbers is intended to assure compatibility among parameters, even when spelling and translations may differ.

The program has been distributed on diskettes since 1995 by the Department of Environmental Sciences and Policy of the Central European University in Budapest. There has been no funding to update the standards of each country and while some countries have voluntarily supplied new data, the system no longer is accurate and, consequently, of very limited use. Attempts to obtain funding from EU Phare, EBRD, or the European Environmental Agency (EEA) for updating design and content have so far been unsuccessful.

10. Conclusions

A web site for the International Commission for the Protection of the Danube River is needed. The proposed design (see above) for an International Commission web site would combine a home page of publicly available information, supported by complementary web sites in each of the Danube countries, with databases on an "Intranet" (i.e. available only through the use of a password) containing detailed water quality and other hydrographic data which the countries might prefer to keep confidential. It would perform many useful functions, not least of which would be to keep a large audience of specialists in government, non-governmental organizations, and research and educational centers, informed about the Commission and its functions. Added to historical and other presentations would be essential facts about the geography, hydro-geology, and other features of the basin and its countries. It would report periodically on the environmental status of basin natural resources and of the nature and effect of Commission decisions to protect them. It would provide access to the Commission via email for additional questions.

The "Intranet" set of databases would be maintained at the discretion of the countries. A network of between 40 and 50 monitoring sites throughout the basin using compatible sampling and testing technology already exists. This network should be expanded and should become the basis for providing the accurate water quality trend data now lacking in the basin. If supported by Commission members this "Intranet" could become an aggressive tool for introducing waste treatment into the former eastern bloc countries.

A simple text system is also required. The cost of a simple text system need not be high. Many of the Danube countries themselves now have the technical ability to create and sustain web sites. The Commission should be able to afford a modest contribution of its own which can borrow from the country sites, if necessary. The headquarters of the Commission will be in Vienna. With a web master and technical help, with modest office space the proposed DIS could be sustained for $35,000 to $40,000 a year and could be implemented by the Commission Directorate.

Contributions of Complementary Data and Information Are Affordable. The Commission members should agree to contribute complementary data and information for use by the central web site. The contact directories, legislation summaries, water quality standards, and information about priority clean-up sites and protected areas, should be provided on web sites sponsored by environmental ministries or other agencies. Costs will vary but within the Eastern European countries a level of about $20,000 would be required (assuming an existing server, office space, and help, and services contributed by government agencies).

Sharing of confidential data will continue to be resisted. Sharing classified data on hydrological phenomena and water quality levels will be continue to be resisted by basin countries because the collection and maintenance of such data is much more costly governments also wish to avoid the risk of publicity about some environmental issues. Trend data for most parameters continues to be unavailable in the basin except for that collected under the Bucharest Declaration. Activation of additional sites has proceeded slowly and the ultimate goal of seventy plus sites has yet to be reached. The cost of work accomplished so far, and estimates of future needs, must be one of the first orders of business for the new Commission Directorate. Support from the European Environment Agency is anticipated and would be helpful.

An environmental standards database for the present and the future is needed. The original cost of the ESD developed in 1993-1994 under the sponsorship of the EBRD, approximately $200,000, paid for the collection of a very large number of standards, for training operators in the CEE countries, for translations, for the design of a system that would accommodate the standards and text material, and for loading data into the system. While technical improvements need to be made in the database (adapting it to Windows 95 or 98, for example), the basic design remains adequate and could be expanded to include environmental standards from other countries. The EEU Department of Environmental Sciences and Policy made important improvements in the operation of the database. However, no funding for updating the contents of the database has been made available by the EBRD, EU Phare, or the EEA, which have expressed interest in its continuation from time to time.

11. Recommendations

Continue the ESD in its present format. Continue the ESD in its present format and plan to support it on the Internet as well as through the distribution (at a modest cost) on diskettes. The CEU server is available. The cost of updating, which requires the time of a researcher, plus some technical support, would not be high, probably no more than $25,000 to $30,000 per year. The EEA should consider taking over the ESD or providing help to it.

Consider a global expansion of the ESD. Serious consideration should be given to making ESD a globally applicable system. The UN Environmental Program should consider appointing a panel of experts to review ESD's possible use on the Internet and its expansion to other parts of the world. The international community of environmental organizations, research and educational institutes, private sector corporations, and government ministry officials should be represented on such a panel. The value of a global environmental standards database should be evaluated by the panel and, if there seems to be need for it, plans should be made to adapt the ESD to other countries and regions.

References

1. McCaffrey, S.C. (1993) Water, Politics, and International Law, in Gleick, P.H. (ed.), *Water in Crisis. A Guide to the World's Fresh Water Resources*, Oxford University Press.,103, endnote #105.

2. Murphy, I.L. (1997) *The Danube: a River Basin in Transition*, Kluwer Adademic Publishers, Dordrecht.

3. Murphy, I.L. (1997) *Protecting Danube Resources: Ensuring Access to Water Quality Data and Information*, Kluwer Adademic Publishers, Dordrecht.

4. Vlachos et al, eds. (1987) *The Management of International River Basin Conflicts: Proceedings of a Workshop Held at the International Institute of Applied Systems Analysis*, Washington, D.C.: George Washington University.

Web Sites

Rhine Commission
(http://www.iksr.org)

Consortium for International Earth Sciences Information Network
(http://www.ciesin.org)

U.S. BUREAU OF RECLAMATION DISTRIBUTED DATABASE APPLICATIONS

JOHN OSTERBERG
D-5500
Bureau of Reclamation
P.O. Box 25007
Denver, CO 80225 USA
Josterberg@do.usbr.gov

Abstract

Today's water managers face a world where competing interests and demand for limited resources have raised public awareness on water management. Environmental concerns, complex allocation and supply issues have required water managers to collect and interpret more data than has been heretofore required historically. The necessity for high quality data has resulted in the need for standardized collection methods, quality assurance/quality control programs, and adequate training. Assessing trends through historical data has demonstrated the need for metadata (data about data) for comparability.

Assessing operation impacts now requires multi-disciplinary teams; the use of distributed databases facilitates this requirement. Distributed databases offer water managers the opportunity to collect and maintain databases throughout the scientific community thus allowing researchers and water management agencies to maintain their databases while sharing access with others. This has resulted in monetary savings since data collection efforts are less likely to be duplicated and limited data collection resources can be fully utilized. Distributed databases have fostered partnerships among various concerns since understanding complex water resource problems usually requires more capability than one water management agency may be able to provide.

Key Words

Agrimet, distributed databases, environment, GCES, GIS, Glen Canyon, GOES, Hydromet, management, metadata, NEPA, quality assurance, STORET, USBR, WATSTORE

T. Naff (ed.),
Data Sharing for International Water Resource Management: Eastern Europe, Russia and the CIS, 229–238.

1. Introduction

The U.S. Bureau of Reclamation (USBR) was founded in 1902 as a civil works construction agency. The agency was staffed principally by engineers, builders, and technicians. Hoover Dam on the Arizona/Nevada border and Grand Coulee Dam in Washington are just two of the better known testaments to the Bureau's abundant engineering and construction accomplishments.

USBR's traditional role was to develop water for economic use in the 17 arid states in the American west. These efforts have made the agency America's largest wholesale water supplier. The agency is the ninth largest electric utility in the United States with 58 hydroelectric power plants, it delivers water to 31 million people each year, and supplies irrigation water to over 10 million farmland acres. Associated with these facilities are 308 recreation sites which receive 80 million visitors annually.

In the 1990's the agency shifted from constructing water projects to managing water because of changing national priorities. This shift was a natural one but it has not been without controversy. Reclamation is effectively managing water through a variety of innovative techniques and in partnership with all interested parties. The agency continues working to improve water resource management, including conservation, water reuse, and efficiency, in an effort to ensure reliable water supplies for the future. The Bureau of Reclamation's goal is to ensure adequate water supplies to meet all the needs of the agricultural, industrial, and municipal sectors of a growing American West, as well as the requirements of Native Americans, fish and wildlife, and recreational users of water in that part of the U.S.

This new role has required USBR to change the way that it conducts business, specifically, how we collect and share data. Historically, agency decision making on basinwide and individual projects was made in-house or with selected entities. This resulted in managing water resources for certain beneficiaries to the exclusion of others. Major decisions on water-related operational changes now involve all interested parties in a public decision process. Two of the primary catalysts for this process have been the National Environmental Policy Act of 1969 (NEPA) and to a more limited extent, the Endangered Species Act of 1972. These legal authorities have had a positive impact on ensuring equal treatment of all project interests.

NEPA requires that all potential positive and negative impacts of a project be addressed. This requirement includes collecting substantial data on the natural and man-made environment to develop an environmental assessment which is an integrated environmental management tool. Developing an environmental assessment requires water managers to collect data on the physical setting, on air quality, water quantity/quality, plants, animals, transportation, social setting, etc. This range of data collection necessitates the development of multidisciplinary teams and accompanying databases to manage and evaluate the impacts.

Environmental and water data collection efforts by the Bureau of Reclamation are either the responsibility of other Government agencies such as the U.S. Geological Survey (USGS) or they are collected within the Bureau. Flow data collected by the USGS is included in the WATSTORE database operated by the USGS, and water quality data is entered into the STORET database maintained by the U.S. Environmental Protection Agency. Biological data collected by USBR is, in most cases, entered into individual state databases. Other information is generally incorporated into databases in the various USBR offices. These distributed databases are available upon request and in some instances are made available electronically through the Internet or via modem.

For purposes of this study, two databases — the Hydromet/Agrimet databases in the Pacific Northwest and a project specific GIS database developed as part of the Glen Canyon Dam Reoperation Studies — have been included that demonstrate the types of data collected by USBR and that also illustrate the kinds of partnerships that are formed by the Bureau in its efforts to collect data. The Glen Canyon Dam is located in Arizona on the Colorado River and is part of the Colorado River Storage Project. Both of these programs are located in watersheds that extend into neighboring Mexico or Canada.

In many instances, USBR has served as a catalyst in the formation of large data networks that not only serve the Bureau's needs, but the needs of other Federal, state, and local entities. Although establishing these shared databases can be difficult, the overall savings for all interested parties warrants the effort in this approach. In establishing such databases, the importance of data quality and standardized methods cannot be stressed enough.

2. Hydromet

The Boise/Minidoka Hydromet System (Hydromet) is a comprehensive data collection system which is of critical importance to the Bureau of Reclamation. The primary use of Hydromet data is to support reservoir and water project operations, water management, and water supply forecasting for the Bureau's multipurpose reservoir systems in the Columbia and Snake River basins. Water uses supported by Hydromet include flood control, irrigation, power generation, water quality, water conservation, fish and wildlife management, research, and recreation. The Hydromet database provides an excellent source of information for water management planning activities.

Hydromet was developed to assist local government and non-government dam operators. Historically, streamflows and unmanned dams were monitored by employees going out to remote areas for visual inspections. Before Hydromet, several days could go by before accurate water data were available for a given location. Telephones were used to coordinate management efforts throughout the river systems. Adequate, timely information was most difficult to obtain during severe weather conditions (such as widespread flooding), just when this information was most needed. A structural failure in the late 1970's at Teton Dam highlighted the importance of having information readily available when rapid decisions become necessary.

A need for better and more timely information in conjunction with improvements in technology led to the establishment of the Hydromet program. Hydromet consists of approximately 300 unmanned data collection platforms located at dams, streams and mountainous areas in the Pacific Northwest, plus computer systems in Boise, Idaho and Yakima, Washington. The system also collects data from approximately 1,100 similar stations maintained by other organizations. These stations typically collect data in 15 or 60 minute increments, then transmit collected data every 4 hours via the Geostationary Operational Earth Satellite (GOES) network to stations where it is needed. Data which indicates abnormal conditions can be transmitted more frequently on special satellite channels. Data are received at the central downlink site in Boise (and at NASA facilities in Wallops Island, Virginia for redundancy). Computer systems in Boise process and distribute the data to system users and organizations.

Other sources of Hydromet data include radio-controlled stations (that can be interrogated at any time) and instrument readings manually entered into the computer system. The relatively short-range, radio-controlled sites are commonly found near large water projects where bi-directional control systems or time-critical data are important. Data from the radio sites are transmitted to the central Boise site via Internet links.

Data collection platforms carry a variety of environmental sensors that monitor water level conditions, gate position, temperature, precipitation, water and barometric pressure, dissolved gas content, solar radiation, evaporation, wind speed and direction. Many data collection platforms exist as a result of cooperative agreements between USBR and other organizations.

In most cases, data provided by the aforementioned platforms are only preliminary in nature. Reporting agencies typically review this data at the end of the year and do not publish official data for up to a year after the end of the water year. Near real-time data meets the management needs for this application, but may not be appropriate for other uses. The Bureau's Hydromet data reports are presented in formats generated by legacy computer systems; as resources permit, these reports will be modified to improve clarity and presentation for future public use.

Hydromet data is shared with other Federal and state agencies, other USBR offices, universities, public utilities, and agricultural users or other interested parties. Hydromet data are exchanged with comparable data from sources that include the U.S. Army Corps of Engineers, U.S. Geological Survey, the Natural Resources Conservation Service, the Bureau of Land Management, the Department of Energy, and Washington State University.

3. Agrimet

Another system related to Hydromet is the Agrimet data collection system. Agrimet is a cooperative effort between USBR, the Bonneville Power Administration, and a number of other organizations in the Pacific Northwest. The focus of the Agrimet program is improved water management including water conservation, and efficiency improvements. Agrimet collects meteorological and other data for calculation of potential evapotranspiration and in support of various other agricultural applications and research. Information on crop water use generated by the Agrimet system can be used as a tool for scheduling crop irrigation. Data is collected, processed, and stored using resources common to both Agrimet and Hydromet systems. Other USBR offices have systems similar to both Hydromet and Agrimet for managing resources within their purview.

The Hydromet and Agrimet databases are available on the Internet. The associated web page provides data and links relevant to water supply conditions in the Pacific Northwest Region of Reclamation encompassing, roughly, the Columbia, Snake, and Rogue River basins. This area includes portions of the states of Washington, Oregon, Idaho, Montana, and Wyoming. Water Storage Reports summarize data from up to 53 reservoirs in this area. Users can generate reports on individual sites or the entire basin. Also available on this site are the Palmer Drought Indices and snowpack data that are extremely important in evaluating the current water supply status in the basin. The Palmer Drought Indices are provided by National Oceanic and Atmospheric Administration (NOAA) and the snowpack information is collected and posted by several agencies.

Funding for the Hydromet and Agrimet databases comes from USBR. The non-USBR stations which encompass the majority of the network are funded by the individual entities and through partnership agreements to share information with the Bureau. Satellite uplink stations and some of the preliminary costs to non-Federal entities in the construction of these stations was funded by the Bonneville Power Administration (BPA), which funded this activity to promote water management. BPA is a Government agency that markets Federal hydropower including that generated by USBR dams. Improved water management has allowed the Federal dams to increase hydropower production. USBR maintains the Hydromet and Agrimet distributed databases and is responsible for maintaining free access to all interested parties.

The Hydromet and Agrimet databases and associated software were developed by USBR for a VAX computer. Since this software has been in service, new more sophisticated software packages have become commercially available. These relational database packages are user friendly and offer more flexibility for the end user. USBR will probably shift to one of these new relational database packages in the future.

The Hydromet and Agrimet systems provide a cost-effective, efficient way to manage the Bureau's water resources because data is virtually real-time and available in adverse conditions that require immediate management. Because of the automated features of these programs, personnel costs are dramatically reduced by eliminating the need for employees routinely to check hundreds of field instruments in remote areas every few days. New stations can be integrated quickly in response to emergency situations such as potential mudslides. The public is better served and public safety is improved since rivers are monitored hourly and not weekly as they were just a decade ago.

The building of Hydromet distributed databases has improved the quality of available data which resulted from initial efforts to develop standards, and to provide technical assistance and training. The Hydromet program has also allowed water managers to look holistically at the river system. Data shortfalls were easily identified, and in some cases funding was made available to fill gaps that would have never been accomplished without the sharing of information.

Making data more readily available has provided some additional and unexpected benefits. In assessing the impacts of USBR's projects on the Snake River Salmon, which is an endangered species, research biologists were using daily flow records to evaluate habitat. When the Hydromet program was established biologists were able to obtain near-time flow records in 15-minute intervals. This high-resolution flow data indicated that for short periods of time critical spawning beds were being exposed. This was a vital finding that may not have been discovered without accessible databases.

4. Glen Canyon Environmental Studies Database (GCES)

Prior to the completion of Glen Canyon Dam in 1963, the Colorado River flowed through the Grand Canyon carrying a heavy load of sediment, and its seasonal water temperature ranged from 40 to 80 degrees Fahrenheit. Annual flooding levels of 100,000 (or more) cubic feet per second (f^3/s) scoured the canyon. Post-dam water, flowing from the depths of Lake Powell, is relieved of its sediment load and cooled to a constant 46 degrees. Depending on cyclical hydropower demands, the Colorado River's flow rate varies from an average minimum of 5,000 f^3/s to an average maximum of 30,000 f^3/s.

Increasing public concern about the impacts on the Colorado River environment of the abrupt and highly variable flows, resulted in the initiation of an environmental impact statement (EIS) and a long-term monitoring program. The study area is located within the Grand Canyon National Park, Glen Canyon National Recreation Area, and Lake Mead National Recreation Area. This is a very sensitive area of national importance that is a major tourist destination. The operations at Glen Canyon Dam are highly visible and the operations of this dam have far a reaching impact on water supply, power production, the environment, and recreational experiences. This situation resulted in the need to collect a vast amount of data, including historical data that would be needed for the determination of trends.

Effective use of such data requires the development of an integrated and spatially correct database to evaluate and monitor impact. The Bureau of Reclamation provided the lead in developing a digital database so it could be updated, analyzed, and displayed through a Geographic Information System (GIS), making it available to other resource agencies. The EIS includes data on air quality, cultural resources, endangered species, fish, hydropower, non-use values, recreation, sediment, vegetation and wildlife habitat, and water resources.

The group from cooperating agencies that USBR put together for the environmental impact assessment included its own personnel and representatives from the Bureau of Indian Affairs, the Environmental Protection Agency, the National Park Service, the U.S. Fish and Wildlife Service, the Western Area Power Administration, the Arizona Game and Fish Department, the Hopi and Hualapai Tribes, the Navajo Nation, and a private consulting firm. This team was responsible for formulating alternative ways to operate the dam and for assessing impacts on the environment. For resources that were to be studied in detail, subteams were formed to make the impact determinations, document their findings, and draft that particular section of the EIS. For the other resources, individuals with expertise in a particular field were assigned the responsibility for determining the impacts and preparing documentation.

The findings of the EIS resulted in operational changes at Glen Canyon Dam. Historically the dam was operated to meet peak power demands in the southwest. This resulted in daily fluctuations from 5,000 f^3/s to a maximum of 30,000 f^3/s and to containment of all spring runoff for power production. The preferred operating alternative substantially reduced the daily fluctuations and ramping rates (rate of increase or decrease in flow through the turbines) below the historic operations and includes annual habitat maintenance and occasional beach building flows. This system conforms more closely with the pre-project hydrograph. The benefits of this release pattern improves wildlife habitat, restores sandy beaches, and enhances recreational opportunities below the dam. However, there was a price paid for these changes: the loss of peak power capacity.

As part of a long-term monitoring effort, a GIS database was proposed as a consolidating tool to enable the integration of data and assessment of the impact of variable flows on Grand Canyon resources. The GIS functions as a visual and analytical tool, as well as an archival device to facilitate data storage, access, and retrieval.

5. Database Development

The GCES/GIS database is a multilevel structure which includes base map and overlay data that ranges from general to specific. There are three tiers of accuracy available within the multi-level structure. Each tier of accuracy is determined either by the scale of the base map or methodology used in the development of the original data set. The first tier consists of data referenced to a 1:24,000 quad sheet; at best, it will reach the National Map Accuracy Standard of 40 feet in the horizontal and half a contour interval in the vertical. The second tier of data is photo referenced and transferred to 1:2,400 orthophoto grid base map developed for this project. The methodology used to create the 1:2,400 level can produce a digital product with a horizontal accuracy of 2.0 meters and a vertical accuracy of 1.0 meter. This second tier of data is being used in the development of the digital products for the 13 monitoring sites. The third tier of data is survey referenced data. This data collection methodology can produce digital data with sub-centimeter accuracy. The GCES office has surveyors on staff to assist contributing scientists in developing new data sets or for geo-referencing historic data with local control.

It was determined that a pilot study should be initiated to develop a methodology prior to committing resources to all monitoring and special study sites. The Nankoweap Canyon area was selected for the pilot project because this area encompassed the following key factors:

- Data collection by multiple agencies
- Extensive quantities of current and historical data
- Cultural significance
- Ecological diversity
- Water quality assessment
- Critical endangered species habit.

Historic data, defined here as "data developed prior to the 1990 GCES/GIS base map," was tested in the pilot study to see if they could be integrated into a valid and usable data set for the GCES/GIS database. The effort to integrate historic data proved that it is possible to fold most geo-referenced data into the GCES/GIS database. The exercise highlighted the need for metadata for both historic and current data sets. Metadata is data that describes the methods used in collection and any other factors that may impact or be relevant to that specific data set. Without metadata it would be virtually impossible to determine the accuracy or validity of data being used in the

GCES/GIS database. The potential size of the GCES/GIS database warrants the use of established metadata standards.

6. Verification and Transfer

River trips were made during the summers of 1990, 1991, and 1992 to map features of the Nankoweap Canyon area in accordance with the classification scheme. The mapping effort was segregated into three trips due to the limited temporal window available during the scientific flows. Bias of the 1991 data, due to physical changes that might have occurred within the monitoring sites between trips, was eliminated by mapping only classes represented on the photographic base map product. Emphasis was on mapping the new and old high water vegetation, but the mapping effort also included other features highlighted in the classification scheme. In order to standardize the study area being mapped, field work for sample sites was done during the scientific low flow (5,000 f^3/s) represented by the CIR photography and base map products.

The GCES Environmental Impact assessment was completed in March 1995. The Glen Canyon Dam's final environmental impact statement recommended that the periodic high flows be used for one to two weeks in the spring to provide habitat maintenance. The purpose of these flows is to rejuvenate backwater channels that are important to fish habitat and maintain sandbars that are important for camping. Habitat maintenance flows differ from beach/habitat flows in that they would occur annually regardless of reservoir levels. The preferred alternative for operating the dam also reduced daily flow fluctuations and provided periodic high steady releases, while allowing limited flexibility in power operations.

7. GCEIS Lessons Learned

Many issues were addressed during the development of the GCES/GIS database. Furthermore, in the course of dealing with the issues, standards were established. Contributing scientists have to be aware of the factors that will influence their data and its integration into the GCES/GIS database. It must be stressed that the methodology of data development depends on the final use of the product. If the data are to be used for long- term monitoring, then accuracy and repeatability are important. Various levels of accuracy necessitate different methodologies for database development. If repeatability is an issue, the data development methodology must be standardized for each monitoring cycle. Metadata must also be in place and will be required with all contributed data.

USBR made every effort in this study to ensure that data collected by all researchers would conform to the GIS database. To ensure compliance the Bureau made the base maps available only after working with the individual researchers on formats and minimum standards. In hindsight, training on the base maps and data standards was well worth the effort since all data overlays, developed in different locations by different entities, are all compatible with the GIS base map. Security of the GIS database was also of great concern on this project. Concerns over potential corruption of the database and the possibility that data could be misrepresented mandated that the database would not be made accessible on the Internet. USBR's policy on the database was that all the information was available to all interested parties with the exception of some provisional data, so data was made available through other outlets upon request.

Monitoring of the operational impact of Glen Canyon Dam on the Grand Canyon will continue for the foreseeable future. However, the GIS base maps that were developed by the Bureau of Reclamation in Denver, Colorado have been turned over to the Glen Canyon Research Monitoring Center in Flagstaff, Arizona. The Flagstaff office is operated by the U. S. Geological Survey. This centralization of some of the data was necessary since the monitoring phase is the responsibility of the USGS, while the initial EIS phase was administrated by the Bureau of Reclamation.

8. Conclusions

Data collection to manage these facilities is of national importance. USBR's role in providing water and power throughout the American west is one of the components for economic growth. Without readily accessible data the management of these resources would be haphazard and wasteful, and the public would not be adequately served and potentially placed at risk.

Good water management requires an extensive data collection network. Water management includes collecting project specific information to manage resources efficiently and to predict and monitor project impact. Measuring project impact should include both natural and man-made environments. It is critical that the data collection network support trend analysis of project related impact.

The sharing of data is essential in solving and managing water resources. Water managers cannot afford to collect detailed information on all potential project impacts. Working in partnership with other entities has allowed all parties to accomplish more with less, and reduce the potential for duplication of effort.

Often overlooked are third party benefits from sharing data. Decision making and problem solving are improved when additional data are available. Information collected by others can be extremely valuable for extending the knowledge of current studies.

The use and application of distributed database systems and networks are essential in water management. USBR's Hydromet program is an example wherein distributed databases improved overall efficiency. The building of partnerships and the communication that was required to bring together all the potential supporters and providers of the Hydromet program have provided cost saving benefits to all interested parties. Cooperation in Hydromet has also improved communications among all the concerned actors and has built the trust among them required to address some of the tough issues facing water managers in the Columbia River Basin.

Quality data is critical to water management. Good data collection programs must include training, standardized collection methods, and quality assurance/quality control programs including, of necessity, the incorporation of metadata. Reproducible, accurate data is essential for all aspects of water management.

Databases need to be well conceived. Consideration of the uses to which a database is to be put is critical to its development. Effective databases need to be flexible and user friendly. Databases need to be developed with security, potential for updating, and maintenance in mind, and long-term financial commitments should be in place prior to development. If new technology or techniques are to be incorporated, a pilot study designed to test them is highly recommended.

9. Summary

The Bureau of Reclamation's mission in water management is extremely complex. The case studies on Hydromet, Agrimet, and Glen Canyon Environmental Studies provide only a snapshot of some of the types of programs and activities in which the Bureau is engaged. The necessity for accurate data collection to complete this mission is absolute. Managing river basins and serving all interested parties requires attention to proper data collection techniques. Decision making that is open for public comment requires that the data which is the foundation for the decision process can withstand public scrutiny.

Databases are critical to sound water management. Developing distributed databases jointly with other entities provides opportunities to strengthen partnerships through the technical interactions involved in establishing database goals, formats, setting standard collection methods, and developing quality assurance/quality control criteria. The use of distributed databases has proven essential in understanding the impacts of projects and in enabling USBR to incorporate complex data in its efforts to maximize its capacity in water management.

THE RIVER MANAGEMENT DECISION SUPPORT SYSTEM (RIMDESS©)

TIM BONDELID
Research Triangle Institute
P.O. Box 12194
Research Triangle Park, NC, 27709 USA

Abstract

The River Management Decision Support System (RIMDESS) is an approach to managing water quality using a computerized data integration and modeling tool. RIMDESS integrates databases on instream water quality, dischargers, pollutant loads, regulatory standards, hydrologic structure, and treatment costs and an array of programs for analyzing these data. Users can evaluate the water quality and economic impacts of management options for controlling emissions from point sources and nonpoint source runoff. This paper briefly describes steps in the approach, applications in ten countries, and lessons learned.

Key Words

basin, Danube, data integration, decision support, GIS, management, Odra, Reach file, rivers, RIMDESS, Syr Darya, water quality

1. Introduction

Since 1991, Research Triangle Institute (RTI) has been developing decision support systems (DSS) for water quality management in Central and Eastern Europe, Central Asia, and, now, the United States. RTI's River Management Decision Support System (RIMDESS©) is an approach to managing water quality that centers on a computerized data integration and modeling tool.

RIMDESS is a PC-based system that integrates databases on instream water quality, dischargers, pollutant loads, regulatory standards, hydrologic structure, treatment costs, and an array of programs for analyzing these data. RIMDESS supports informed decision-making regarding the control of emissions in a watershed. Most

T. Naff (ed.),
Data Sharing for International Water Resource Management: Eastern Europe, Russia and the CIS, 239–246.

water resources agencies in the U.S. and abroad have scattered databases and models that do not talk with each other and lack linkages and tools to evaluate the complex interactions in a river basin. RIMDESS allows users to evaluate the water quality and economic impacts of management options for controlling emissions from industrial and municipal point sources and nonpoint source runoff.

The U.S. Agency for International Development (USAID) has sponsored the development and institutionalization of RIMDESS/Danube in the Danubian countries of Bulgaria, Hungary, Romania, and Slovakia since the winter of 1991. Similar assistance to Poland, RIMDESS/Odra, began in 1993 and involves the Upper Odra river basin. In 1996, under the USAID Environmental Policy and Technology Project, RTI applied RIMDESS/Syr Darya in the Central Asian countries of Kazakhstan, Uzbekistan, Kyrgyzstan, and Tajikistan.

RIMDESS started with experience gained in the United States, particularly lessons learned from the success of the U.S. Environmental Protection Agency's STORET and River Reach File. STORET is the U.S. national storage and retrieval system of water quality data, and the River Reach File provides digital hydrologic information to facilitate mapping and development of hydrologic routing models.

2. Key Components of the System

This explanation of the three key components of the RIMDESS system also highlights how to build such a system and the major technical and institutional issues to address. The following will give an overview of each component, a discussion of the technical and institutional issues that should be addressed and a brief description of how it was implemented in the Central Asian version of RIMDESS. The institutional and technical steps are a very general "how to" guide for implementing RIMDESS or a similar system.

The order in which these components are listed does *not* imply the order in which the steps are implemented. The process is such that each component parallels another. For instance, questions about the data management cannot be completely answered until the questions are defined in the analytical steps.

2.1. A DATA MANAGEMENT SYSTEM

The core of RIMDESS is the Reach File, an electronic river network that encompasses the main stem, major tributaries, points of confluence, and flow directions of a river. The Reach File enables the system's other components to predict the downstream effects of complex upstream events.

Many data sets are linked to the Reach File — stream flows, water quality, water intakes, wastewater discharges, and all other information required for a particular user's needs. The data structures used in RIMDESS interact efficiently with many different models, providing a uniquely flexible system.

2.1.1. *Institutional Steps*

- Identify the organizations that "own" or have regular access to the major data sources
- Endeavor to integrate those organizations into the system development process
- Reach out to other organizations that may need further access to this information because they may become future users of the system.

(In Central Asia, technical "homes" were identified in each of the four countries involved; in some cases there was more than one "home" because data existed in different organizations.)

2.1.2. *Technical steps*

- Define the data that needs to be in the system. This is done on the basis of the questions for which the system needs answers (as described in Step 1 in the Integrated Analytical System component below; many types of data are needed no matter what the question)
- Develop the common database structures that can contain the variety of data needed (this has already been done in RIMDESS)
- Develop a set of common nomenclature tables, especially the Reach File river network, that can cover the entire area
- Integrate the data locations onto the Reach File (a variety of methods are available)
- Obtain electronic versions of the necessary raw data — streamflows, water quality, etc. — in the format used by each country involved.
- Set up and run procedures to convert data from each country's formats to the common RIMDESS formats (this step has been successful in over 10 different national-level databases).

In Central Asia, a common Reach File river network was built from a subset of the 1:2,000,000 Digital Chart of the World using Geographic Information System (GIS) technology. The locations of gauging stations, dischargers, water intakes, etc. were hand-marked on maps and then digitized using GIS. Software was written as part of the implementation process to make the Reach File a fully routable network for analyses. Almost all of the data were paper-copy only; our technical "homes" entered the data into computer spreadsheet-like formats that are very similar to their paper tables. Special "middleware" programs were written that convert these spreadsheets into the common RIMDESS formats.

2.2. AN INTEGRATED ANALYTICAL SYSTEM

RIMDESS is built as a flexible, modular, "plug-and-play" system. Each application employs a combination of public domain and custom-built models to address a particular user's needs. Using the data management system to provide common input data for all models and to exchange outputs among models, the user can develop a truly integrated, multi-layered analysis.

2.2.1. *Institutional Steps*

- Define the questions that the system needs to answer (this can be the most difficult step)
- Identify the organizations that "own" the models or tools needed; obtain access to and/or cooperation from them.

2.2.2. *Technical Steps*

- Based on the questions identified as requiring answers, design the analysis sequence necessary to arrive at the answers
- Identify any major weaknesses in the data or analysis capabilities; the question may need to be re-phrased based on current capabilities)
- Assemble the necessary models and analytical tools
- Develop "middleware" that either runs the model within the RIMDESS software or gives the model external accesses to RIMDESS.

In Central Asia, we identified institutional "homes" in each country that were interested in certain questions, and our primary user (USAID) helped define regional-scale issues. There were particular challenges in this implementation because of regional-level policy questions related to water and wastewater pricing plus local-level issues related to examining specific rivers and polluters. We assembled a variety of models and tools, some from previous versions of RIMDESS; one model, a regional mineralization model, was developed locally. All of these models were implemented within the RIMDESS source code and then provided to each country's homes.

2.3. AN EXECUTIVE INTERFACE

Although highly skilled users are able to change the data and models that comprise the operating system, many users and decision-makers want only to understand the effects of changes in the cases of certain input variables. To support this need, RIMDESS applications include a custom-built interface to guide users through the data analysis functions and to accumulate results across a range of conditions. This interface helps users generate summary graphs and tables that are critical for presenting results in a succinct, easily understandable format.

2.3.1. *Institutional Steps*

- Identify the users of the Executive Interface (higher-level decision-makers and managers)
- Meet regularly with these users to demonstrate the Executive Interface and get their feedback for improvements
- Set up training workshops for achieving these objectives.

In Central Asia, a series of workshops and training sessions were conducted, bringing the technical and management staff together to work with the Interface as a commonly understood platform for the system. Regional and national training workshops have always proven to be an excellent way to promote development and use of the system.

2.3.2. *Technical Steps*

- Design a clear, graphically-oriented presentation that addresses the questions integral to the Executive Interface
- Select a software platform that can provide good graphics and a simple user interface
- Integrate the databases, analytical inputs, and results into the Interface
- Focus repeatedly on the look of the Interface in accordance with user feedback.

In Central Asia, a new way (at least for RIMDESS) was designed to present results in which the user is carried through a flowchart of the many different regional-scale analyses that were implemented. For local-level analyses an interface first developed in the Bulgarian version of RIMDESS was adapted.

3. RIMDESS Applications:

3.1. THE DANUBE BASIN — HUNGARY, ROMANIA, SLOVAKIA, AND BULGARIA

The first incarnation of RIMDESS was called DEMDESS an acronym for the Danube Emissions Management and Decision Support System. DEMDESS was developed between 1991 and 1994, and was used to identify opportunities for cost-effective investments in new and enhanced wastewater treatment plants (WWTP) on four tributaries to the Danube. The DEMDESS project two major results: the system was used to support the selection of three WWTP projects which are now being built with GEF funding, and the system has been adopted for national water data management in Bulgaria.

3.2. THE UPPER ODRA BASIN — POLAND

The second application of RIMDESS was developed for the Wroclaw Regional Water Authority (RZGW) in Poland. RZGW has recently been delegated authority from the national government for planning water quality management programs on the Upper Odra River, including investing in new wastewater treatment capacity. The principal outcome of the RIMDESS/Upper Odra has been its use as a part of its routine operations to develop and evaluate proposals for new wastewater treatment plants.

3.3. THE SYR DARYA BASIN — KRYGYZSTAN, KAZAKHSTAN, TAJIKISTAN, UZBEKISTAN

RIMDESS/Syr Darya was developed to assist technical and policy institutions in these four countries to develop water quality management plans, evaluate proposals for increasing water prices, and contribute to regional negotiations on water resource management. This effort produced two key results: First, a precedent and mechanism for sharing water data among the four riparians in the Syr Darya basin were established. Second, technical and policy homes for system maintenance and development were organized in each country in addition to a regional home within the structure of the International Council on the Aral Sea (ICAS).

3.4. THE NEUSE RIVER BASIN — NORTH CAROLINA, USA

RIMDESS/Neuse is used as an instrument for evaluating nutrient loadings into the Neuse River and its estuary, including the Pamlico Sound in North Carolina. The system will be used by the State to design a coordinated program of controls on point and non-point sources to achieve a 30% reduction in nutrient loadings to the river.

RIMDESS is a cumulative system in that each application builds on previous applications. For instance, the RIMDESS water quality model was originally developed for the RIMDESS/Odra version, and has now been incorporated into the RIMDESS/Syr Darya system. The RIMDESS/Neuse system is expanding the decision making process, especially in terms of evaluating nonpoint (dispersed) pollution sources. The diagram shows the full "decision tree" for the new RIMDESS/Neuse, with several new analytical components included.

4. Lessons Learned

From its experience with decision support systems (DSSs), RTI has learned five major lessons that are applicable to watershed planning and management programs:

Place equal emphasis on institutional and technical issues. The best approach is to treat the institutional and technical factors in parallel with one another, allowing the technical solutions support institutional considerations.

Build the DSS "on top" of the existing routine administrative systems. There are three "layers" of RIMDESS functions. The first layer is the central databases, drawn from the existing systems and models. The second layer is integrated applications, e.g., an Emissions Policy Model and a Scenario Manager. The third layer is called "Executive Interfaces," which provide concise, clear graphical presentations for decision-makers, non-technical stakeholders, etc.

Use a data-based approach to integration as opposed to a software-based approach. Modern database management systems have tremendous power for quickly and efficiently integrating a wide variety of data. A data-based approach means that the core of the system is a set of databases surrounded by various individual data systems and models. The individual systems are linked to the core databases through customized "gateways." This means that the DSS is composed of many diverse, individual data and software systems tied together by one DSS data system.

Careful design of central organizing elements is critical. One very powerful element is the Reach File. A Reach File has been built and used to integrate the water data in every country in which RIMDESS is implemented. Other central organizing structures are important too, such as a single unified water quality parameter code table, unified industrial sector codes, etc.

Include costs financing, and economics as integral components of planning, preferably at the beginning of the project. By working financial factors into the system from the beginning, dynamic interactions among water quality changes, costs, and economic benefits can be achieved.

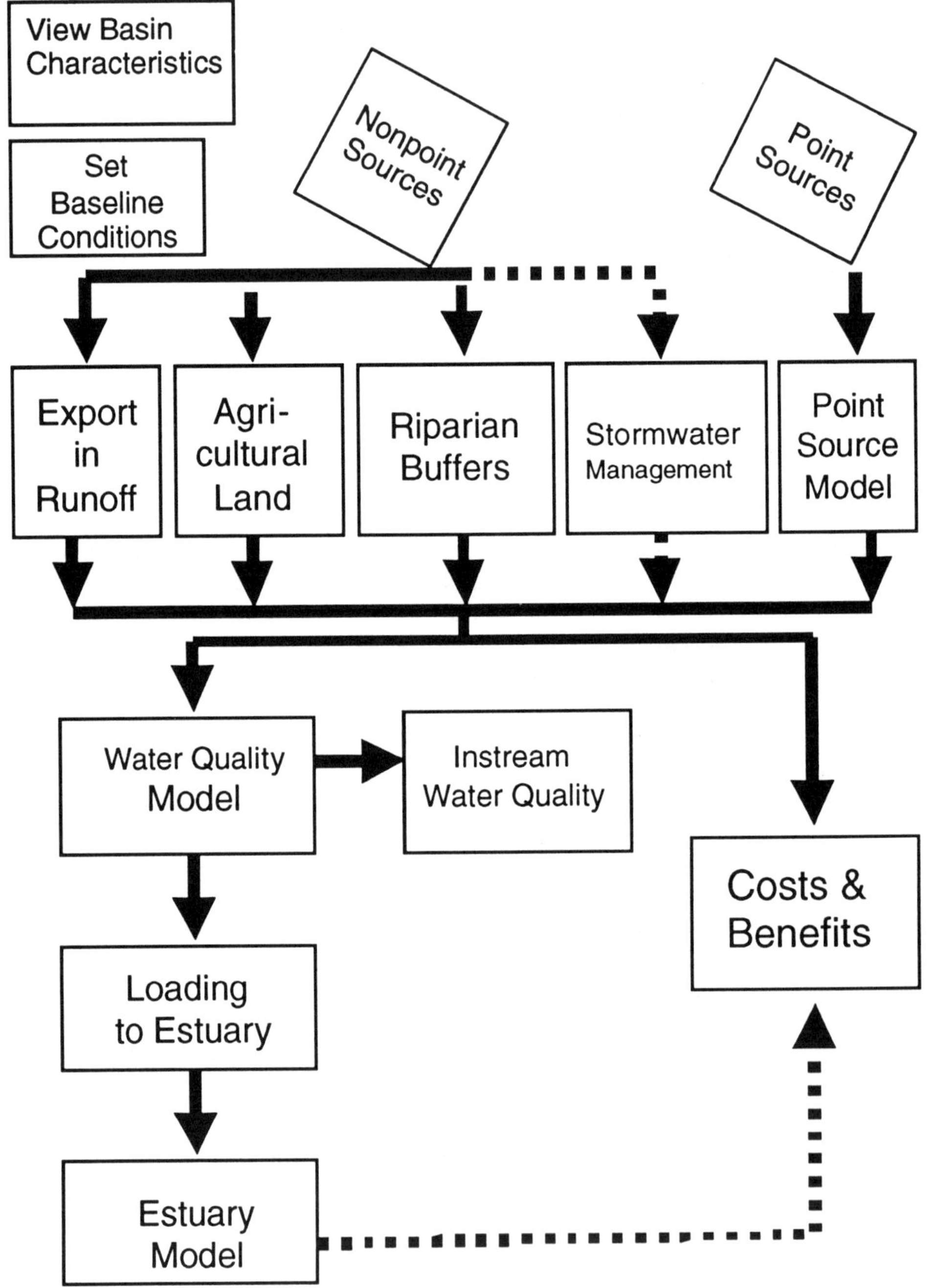
View Basin Characteristics
Set Baseline Conditions
Nonpoint Sources
Point Sources
Export in Runoff
Agri- cultural Land
Riparian Buffers
Stormwater Management
Point Source Model
Water Quality Model
Instream Water Quality
Costs & Benefits
Loading to Estuary
Estuary Model

LIST OF AUTHORS

U.D. ANANICHEVA AND K.S. LOSEV
Institute of Scientific and Technical Information
and the Institute of Geography
Staromonetny, 29
Moscow 109017 RUSSIA

L. YA. ANISCHENKO, Cand. Tech. Sc.
Ukraine Scientific Research Institute of Ecological Problems
6 Bakulina Street
310166 Khariv UKRAINE

N.H. BATJES
International Soil Reference and Information Centre (ISRIC)
P.O. Box 353, 6700 AJ Wageningen, NETHERLANDS

TIM BONDELID
Research Triangle Institute
Research Triangle Park
NC, 27709 USA

BORIS FASHCHEVSKY
University of Modern Knowledge
Kalinovskogo Street. 48-49
220086 Minsk BELARUS

JEAN-MARC FAURES
Water Resources Officer
Land and Water Development Division
FAO
00100 Rome ITALY
jeanmarc.faures@fao.org

T. Naff (ed.),
Data Sharing for International Water Resource Management: Eastern Europe, Russia and the CIS, 247–249.

EDUARD GRANOVSKY *et al*
Kazak Research Institute of Scientific and Technical Information
221, Bogenbai, Batyr str.
Almaty 480096 KAZAKHSTAN
eduard@gran.ksisti.alma-ata.su

ANATOLY V. GRITSENKO
Ukraine Scientific Research Institute of Ecological Problems, USRIEP
6 Bakulina Street
310166 Khariv UKRAINE

MARCELL KNOLMAR AND A. DELI
Budapest Technical University and National Water Authority
H-1111 Budapest HUNGARY
knolmar@vcst.bme.hu

V. P. KROHMAL and A.A. VECHER
Deputy Chief Engineer and Chief, Computer Center
Institute "Turkmengiprovodhoz", TURKMENISTAN

ANNA KWIATKOWSKA AND ANDRZEJ KRASZEWSKI
Warsaw University of Technology
Institute of Environmental Engineering Systems
Nowowiejska 20 00-653, Wasaw, POLAND
anka@iispw.edu.pl; aak@iispw.edu.pl

FERENC LASZLO
Water Resources Research Centre (VITUKI)
H-1453 Budapest, P.O.B. 27, HUNGARY

ARBEN MEMO
Institut Fur Anorganische Chemi
Engsbachstrasse 56, App 222
Siegen 57076 GERMANY
Arben.memo@student.uni-siegen.de

J. S. MINAS
Office of Computing Services
Drexel University
Philadelphia, PA 19104 USA
jminas@drexel.edu

IRENE LYONS MURPHY, PH.D.
Colorado State University
2005 37th Street, N.W.
Washington, D.C. 20007 USA
imurph@aol.com

THOMAS NAFF
Professor, University of Pennsylvania
847 Williams Hall
Phildelphia, PA 19104-6305 USA
tnaff@sas.upenn.edu

JOHN OSTERBERG
D-5500
Bureau of Reclamation
P.O. Box 25007
Denver, CO 80225 USA
josterberg@do.usbr.gov

RICHARD W. PAULSON
U.S. National Weather Service
1325 East-West Highway
Silver Spring, Maryland, 20910 USA

WILLIAM J. SHAMPINE
U.S. Geological Survey
P.O. Box 25046, MS401
Denver Federal Center
Lakewood, CO 08225 USA

ANDRAS TARNOY
Unibox
1026 Budapest
Hidàsz utca 15 HUNGARY

INDEX